TWO-PERSON GAME THEORY

Ann Arbor
Science
Library

Two-Person Game Theory
The Essential Ideas

by Anatol Rapoport

ANN ARBOR
THE UNIVERSITY OF MICHIGAN PRESS

Third printing 1970

ISBN 0-472-05015-X
Library of Congress Catalog Card No. 66-11082
Published in the United States of America by
The University of Michigan Press and simultaneously
in Don Mills, Canada, by Longmans Canada Limited
Manufactured in the United States of America

Preface

Like most branches of mathematics, game theory has its roots in certain problems abstracted from life situations. The situations are those which involve the necessity of making decisions when the outcomes will be affected by two or more decision-makers. Typically the decision-makers' preferences are not in agreement with each other. In short, game theory deals with decisions in conflict situations.

A key word in what has just been said is *abstracted.* It implies that only the essential aspects of a situation are discussed in game theory rather than the entire situation with its peculiarities, ambiguities, and subtleties. If, however, the game theoretician is asked "What *are* the essential aspects of decisions in conflict situations?" his only honest answer can be "Those which I have abstracted." To claim more would be similar to maintaining that the essential aspect of all circular objects is their circularity. This may be so for the geometer but not for someone who distinguishes coins from buttons and phonograph records from camera apertures.

To be sure, the geometer deals not with "circular objects" but with circles. That is to say, the conceptual act of abstracting circularity from all circular objects was performed long enough ago to have been institutionalized in our language and in our science. Hence the geometer can assume that people who wish to study the geometric properties of circles will easily forget all the other properties of circular objects, such as their color, the material from which they are made, or the uses to which they are put.

The game theoretician is in a more difficult position. The aspects of decisions in conflict situations which he considers to be essential are not immediately evident to the mind as is the circularity of circular objects. And even after those aspects are brought into focus, they do not easily stay in focus. We are more emotionally involved with conflicts than with shapes of objects; and so the aspects which happen to be important to us keep intruding into our conceptions of conflict.

For this reason and, of course, also because the subject is new, there is still little understanding of what game theory is and what it is not, of what it could become and of what it cannot become because of certain inherent (not merely circumstance-imposed) limitations. On the other hand interest in game theory as a "science of rational conflict" is extremely widespread in our age of competition, strategy, and gamesmanship. This interest is shared also by the "hard" behavioral scientists, ready to welcome a rigorous mathematically oriented conceptual framework. This combination of lively interest and lack of sufficient acquaintance with the essential ideas of game theory has frequently led to regrettable misunderstandings and confusion; for example, about the uses and misuses of game theory in policy making, and about the relevance of game theory to the social sciences.

This book is an attempt to introduce the theory of games to those interested in it in a way which would bring the essentials of the theory into the focus of attention and keep them there. I suppose a disposition to read this book is sufficient evidence of interest; so I have no misgivings on that score. On the other hand, the question of what to presuppose about the reader's mathematical background is a difficult one. On formal grounds it is possible to maintain that hardly any knowledge of mathematics is required for understanding the essential ideas of game theory. If by knowledge of mathematics, one means an acquaintance with geometry, the calculus, and so forth, then it indeed appears to be true that very little of such knowledge is required beyond the ability to follow the process of solving some quite simple algebraic equations. However, the issue is not technical mathematical knowledge but rather mathematical background, that is, certain habits of thought usually acquired only through the study of mathematics. Similarly, it is not the ability to play an instrument that is required in order to follow the development of a musical thought, say in a symphony, but rather "musicality," certain habits of listening. The ability to think mathematically is like the ability to listen musically. Some of this ability may be inborn; some may be acquired without technical training; and much of it comes with technical training. Therefore it is not to be denied that readers with some mathematical background will follow the ideas of game theory more easily than others. Clearly, however, this book is not addressed to mathematicians, who, if they wish to acquaint themselves with game theory, will turn to the standard treatises. The book is meant as a popular exposition of the subject, which (hopefully) penetrates it in depth.

The difficulty which the nonmathematician experiences in reading mathematical works stems from two sources: a lack of experience with mathematical ideas, and a lack of experience with mathematical *notation*. In the case of game theory, the mathematical ideas are rather simple, their major source being set theory, which is almost en-

tirely self-contained and requires no "prerequisites." *
However the *notation* of set theory is special and thus unfamiliar to readers whose contact with mathematics has been through conventional elementary courses. I believe it is this special notation which makes difficulties for nonmathematicians who wish to familiarize themselves with game theory. Thus, I am told by my nonmathematical colleagues that they find *Games and Decisions* by R. D. Luce and H. Raiffa difficult to read. In my opinion, that book has achieved a triumph in its lucid exposition of the essential ideas of game theory. I can conclude only that the difficulties stem from the fact that nonmathematical readers shy away from unfamiliar mathematical notation which is still typographically prominent in *Games and Decisions,* in spite of the vast simplification over the original entirely uninhibited notation of the fundamental treatise (*Theory of Games and Economic Behavior* by J. von Neumann and O. Morgenstern). On the other hand, J. D. Williams' popular exposition, *The Compleat Strategyst,* from which difficult notation was eliminated, confines itself almost exclusively to the two-person zero-sum game. The idea embodied in this type of game is, perhaps, the foundation of game theory; but one cannot get an idea of a building by examining just the foundation.

My original aim was to present the essential ideas of game theory (including its most interesting and challenging departures from the two-person zero-sum game) for the general reader and for the social scientist, using the barest minimum of mathematical notation. I found this possible in the context of the two-person game; but I still have not succeeded in making an acceptable "translation" of the N-person game theory. Therefore, I have ventured

* We may point out that the so-called "new math" now introduced to American children in grade schools places major emphasis on set theory. The approach has been, as far as I know, a definite pedagogical success.

to offer the present book first, hoping eventually to continue with *Essential Ideas of the N-person Game,* when and if the notation problem is solved.

In the present volume the reader will find only the mathematics of high school algebra and of very elementary analytic geometry, except for an occasional derivative. The only game-theoretical notation retained was that of the game matrix, which is quite easy to grasp and which is, at any rate, indispensable. In addition to the standard topics in the two-person game, a discussion of gaming theory is included, which, in my opinion, is an important link between abstract game theory and an experimentally oriented behavioral science. Specific applications to social science have not been stressed (these being discussed at length elsewhere); but the methodological relations between game theory, decision theory, and social science are emphasized throughout. In contrast to the purely logical and mathematical ideas of game theory, the methodological ideas are controversial. I hope that the dividing line between facts and opinions has been made sufficiently clear.

I am indebted to Prof. Robert M. Thrall, of The University of Michigan and to Mrs. Claire Adler for helpful editorial suggestions; to my colleagues at the Mental Health Research Institute, and to The University of Michigan Press, for encouraging me to write this book. Whatever errors may have crept in despite the help I have received are, of course, my own.

Anatol Rapoport

Contents

1. *Games*

Game theory is to games of strategy what probability theory is to games of chance. And just as probability theory far transcends its role as the logical basis of rational gambling, so does game theory transcend its original guise as the logical basis of parlor games.

Yet the history of game theory is not a replica of the history of probability theory. The birth of the latter is reckoned from some problems posed by the intellectually inclined gambler Chevalier de Méré to the philosopher-mathematician Blaise Pascal. Probability theory received its original impetus from attempts to solve concrete problems.

This was not the case with game theory. No one came to John von Neumann with questions about how to play chess or poker. Nor do we have any evidence that von Neumann, who laid the foundations of game theory practically single-handed, was an outstanding expert in any game of strategy. Nor is there any more reason to expect such skill from a game theoretician than to expect instrumental virtuosity from an orchestra conductor or mechanical ingenuity from a physicist. Game theory is concerned not with any particular game but with all of them, not with technical but with theoretical matters. "What is the best way to play Chess?" is not a game-theoretical question. On the other hand, "Is there a best way to play Chess?" is a game-theoretical question.

It may seem at first that the question is a foolish one. One might be convinced that of course there is a best way to play Chess since there is a hierarchy of Chess players. Some players are obviously better than others; therefore they must use better ways of playing, and so there must be a best way to play Chess.

In response to this line of reasoning, a game theoretician would reply as follows. Yes, there is a best way to play Chess, but the supporting arguments just given for this conclusion are not valid. It is not true in general that if we have a number of quantities clearly comparable with respect to magnitude, then one of them must be the largest. Consider, for example, the sequence of natural numbers, 1, 2, 3, etc. Each is larger than the preceding; yet there is no greatest natural number. Nor must the magnitudes increase without bound so that an ordered set of them will have no largest member. Consider, for example, the set of all real numbers between zero and one, exclusive of zero and one. This is an ordered set, because of any two such numbers one is definitely the larger. Yet the set as defined has neither a smallest nor a largest number. To be sure, it is true that although the magnitudes of the numbers from zero to one are not increasing without bound, the quantity of such numbers is infinite; and it is true that if a set ordered by magnitude has only a finite number of members, there must be a smallest and a largest. Therefore, the question of whether there is a best way to play Chess is related to the question of whether the number of ways to play a game of Chess is finite or infinite. This number happens to be finite. But whether it is an ordered set in the sense that it is always possible to decide which is the better way of playing between any two, depends on the principle according to which the ordering is made. Finally, it does not follow that if there is a best way of playing a game, the player who knows it will always win. For the player who plays against him may also know a "best way." What happens when both have this knowledge is also an open question. In some games the victory must go to the one or to the other; in other games, if both know a "best way to play," the outcome must always be a draw. In still other games, not necessarily games of chance as these are usually understood, the outcome will nevertheless depend on chance.

All the matters just touched upon pertain to the theory of games. They are not matters related to any specific game; they are matters related to games in general. The way the questions have been put suggests that games can be classified according to the way the questions are answered.

Game theory is, accordingly, very largely concerned with the classification of games, and in this it has much in common with other sciences which at a certain stage of their development were concerned mainly with classification.

For example, biology was for many centuries mainly a classification science (a taxonomy). Biologists sought a "proper" way to classify plants and animals. It would seem at first that what the "proper" principles of classification are depends crucially on what the classifier is interested in. For instance, someone coming in frequent contact with animals in a primitive life environment might classify animals into large and small, or into dangerous and harmless, or into edible and inedible. There comes a time, however, when observation and description of nature becomes more or less separated from immediate utilitarian interests. Accordingly, biologists soon recognized that although mice and lizards were both small animals while horses and crocodiles were both large animals, nevertheless mice were more closely related to horses than to lizards while crocodiles were more closely related to lizards than to horses.

Formal Decision Theory

The principle according to which game theory classifies games is best understood if game theory is viewed as the branch of mathematics concerned with the formal aspect of rational decision. The emphasis is on the word "formal," which in this context means "devoid of content." As has been said, game theory has been hitherto developed as a branch of mathematics. Mathematics treats of formal relations devoid of content. For example, arithmetic is not concerned with apples, sheep, or dollars, but only with relations among numbers be they of apples, sheep, dollars, or divorces. Geometry is not concerned with land tracts or shapes of objects but only with spatial relationships. Similarly a mathematical theory of rational decision is concerned not with the problem of making wise decisions but with the logical structure of problems which arise in connection with the necessity of making decisions.

A decision problem is the problem of choosing among a set of alternative actions. Clearly such a problem is meaningful only if whoever must make the choice has some idea of what the consequences of his choice may be. In the simplest case, each choice of action leads to a single specific consequence, so that the choice of action is equivalent to a choice of a consequence among a set of consequences. Hence the first condition which must be fulfilled if a decision problem is to have meaning is that choices have known consequences.

Next, choice is meaningful only if the chooser has preferences. Thus the second condition which must be fulfilled, if the decision problem is to have meaning, is the existence of preferences of the chooser. Again we may consider the simplest case in which all the possible consequences can be clearly rank ordered, calling the most preferred consequence first choice, the next most preferred second choice, etc. We have already assumed that in the simplest case, each choice of action has exactly one consequence, known to the chooser. Therefore the decision problem in this case reduces to the problem of assigning preference rank orders to the consequences, noting which choice leads to the most preferred outcome, and choosing it.

Most decision problems are not so simple. As a rule, an action may lead to a number of different outcomes, and which one will actually obtain in a given instance is not known to the chooser. The structure of the decision problem depends critically on the factors that determine which of the possible outcomes of an action will actually occur. If the actual outcome is determined purely by chance, we have a decision problem under risk or uncertainty. Probability theory is sometimes a valuable tool in such decision problems.

In some cases the outcome of one's choice of action will be determined not by chance but by someone else's choice of action. These situations fall properly within the scope of game theory. The classification of games is guided by the sort of decision problems that arise in the course of a game. Moreover, any situation having the abstract (formal) features of a decision problem of the same sort that appear in a game can also be called a game.

In short, what distinguishes games from nongames from the point of view of game theory is not the seriousness or lack of seriousness of a situation, nor the attitudes of the participants, nor the nature of the acts and of the outcomes, but whether certain choices of actions and certain outcomes can be unambiguously defined, whether the consequences of joint choices can be precisely specified, and whether the choosers have distinct preferences among the outcomes.

It may seem surprising that nothing has yet been said about the *rules* which define a game. Rules are important only to the extent that they allow the outcomes resulting from the choices of the participants to be unambiguously specified. Once these choices have been listed and the outcomes resulting from the participants' choices have been ordered according to the preference of each participant, the rules according to which the game is played are no longer of any consequence. Any other game with possibly quite different rules but leading to the same relations among the choices and the outcomes is considered equivalent to the game in question. In short, game theory is concerned with rules only to the extent that the rules help define the choice situation and the outcomes associated with the choices. Otherwise the rules of games play no part in game theory. How this comes about will, we hope, become clear in the next chapter.

The Essential Features of a Game

Game theory is concerned with situations which have the following features.

1. There must be at least two players.

2. The game begins by one or more of the players making a choice among a number of specified alternatives. In ordinary parlance such a choice is called a move. In game theory "move" refers rather to the situation in which the choice is made, for example, the specification of who is to make the choice and what alternatives are open to him. Thus the game of Chess begins with a range of twenty alternatives open to the player called White. [These twenty alternatives are advances of one or two squares for each of the eight pawns (sixteen alternatives) and two squares open to each of two knights (four alternatives).]

3. After the choice associated with the first move is made, a certain situation results. This situation deter-

mines who is to make the next choice and also what alternatives are open to him. For example, in Chess, the players alternate in making their choices regardless of the situation. In many card games, however, the player to make the next choice is always the player who took the last trick. In all games, however, it is clear in every situation which player is to make the next choice and what alternatives are open. In Chess, after White has moved, Black has the same twenty alternatives regardless of how White has moved. But this is not the case on the next move (White's). For example, if White advanced his King's pawn two squares on his first move, and Black responded by advancing his King's pawn two squares, then White cannot advance the King's pawn on his next move. Had Black responded in any other way, White would have that alternative open.

4. The choices made by the players may or may not become known. In Chess all the choices are known to both players. But in a variant of Chess called Kriegspiel none of the choices made by one player are made known to the other (although these choices can sometimes be inferred from other information). This circumstance makes this game quite different from Chess. Games in which all the choices of all the players are known to everyone as soon as they are made are called *games of perfect information.* Chess, Checkers, Go, Backgammon, and Tic-Tac-Toe are all games of perfect information. Most card games are not games of perfect information. It is instructive to see why this is so. In most card games, the cards are dealt face down, so that each player can see only his own hand. Imagine that Chance is a fictitious player in such a card game. The first move is Chance's. She is to "choose" between all the possible arrangements of the deck. The result of this move is unknown to the other players. Thereafter, the results of each move may be known to everyone (e.g., if the cards are played face up), but Chance's choice remains un-

known at least for some time. (It may be inferred later.) Therefore such a card game is not a game of perfect information. The importance of singling out games of perfect information is that in such games there is always a "best way to play" which can be specified without mentioning chance, while in other games this is not necessarily the case.[1]

5. If a game is described in terms of successive choices (moves), there is a *termination rule*. Each choice made by a player determines a certain situation. For example, following each move, the arrangement of the pieces on the chessboard defines a situation. Certain situations in Chess are called "checkmate." When such a situation occurs, the game is ended. Many card games end when all cards in the deck have been played; a game of Tic-Tac-Toe ends when one player has three naughts (or crosses) in a straight line, etc. While in common parlance one speaks of the end of a game, in game theory one speaks of the end of a *play* of the game. The word "game" is reserved for the totality of rules which define it. We shall for the most part adhere to the usage of game theory, but on occasions, when there is little danger of confusion, we shall revert to common usage in order to avoid awkward expressions. Thus we shall say, "Three games of Chess were played" instead of "Three plays of Chess were played."

6. Every play of a game ends in a certain situation (i.e., one of the situations which define the end of a play). Each of these situations determines a payoff to each bona fide player. A bona fide player is one who (1) makes choices and (2) receives payoffs. Thus, although we called Chance a player in a card game (because she "chose" the arrangement of the cards), we cannot call her a bona fide player, because she gets no payoffs. In a game, as defined in game theory, there must be at least two bona fide players, in the sense that each of them makes choices and receives payoffs.

Solitaire is not a game from the point of view of game theory. The player of Solitaire is, to be sure, a bona fide player, because he both makes choices (in the more sophisticated forms of Solitaire) and receives payoffs (at least wins or loses). However, the other player, Chance, only makes choices (arranging the cards), but receives no payoffs. Chance is only a dummy player. Nor can the House (if solitaire is played in a gambling establishment) be called a bona fide player, because although the House receives payoffs, it makes no choices. Playing the slot machine is not a game either. Curiously, the slot machine can be considered a bona fide player. It makes choices (to be sure at random, but this does not matter) and receives payoffs. But the person who plays the slot machine only receives payoffs. He makes no choices, because the only thing he can do in each play of the game is insert a coin and pull the lever. Therefore he is not a bona fide player. He is only a dummy player.

If the above six criteria are satisfied, we can speak of a game.[2] A particular game is defined when the choices open to the players in each situation, the situations defining the end of a play, and the payoffs associated with each play-terminating situation have been specified.

2. *Utilities*

So far nothing has been said about the nature of the payoffs. In game theory it is simply assumed that numbers, positive or negative, can be specified as payoffs for each of the terminating situations. In card games it is quite natural to view the money gains and losses as the payoffs. Some games like Chess or Tic-Tac-Toe are not usually played for money. However, the outcomes of these games are clearly defined as "win," "draw," or "lose."

It is assumed in game theory that all outcomes can be interpreted as numbers. For example, if the only distinguishable outcomes are win, lose, and draw, 1 can stand for win, −1 for lose, 0 for draw. *How* these numbers are assigned is not the game-theoretician's concern. It is conceivable, for example, that a Chess player would rather risk losing a game than settle for a draw. It is even conceivable that a man playing Checkers with a child would rather lose than win. In that case a larger payoff must be assigned to his loss than to his win.

If payoffs are in money it is quite likely that the psychological "worths" of the amounts of money do not correspond to their numerical values. To win $20 in a poker game may be "worth," to a given Poker player, more than twice as much or less than twice as much as to win \$10. Since the magnitudes of the payoffs play a part in defining a particular game, the game remains undefined if we do not know what payoff magnitudes are assigned by the players to the outcomes, even if the latter are specified in terms of monetary payoffs. However, this problem is bypassed by the game theoretician, who assumes that the payoffs are given. Once the payoffs are given, the game is defined. Once the game is defined, game-theoretical analysis can be brought to bear on it.

The assumption that something is "given" is not as crippling to a theory as might first appear. Every theory must begin with some givens. For example, the mathematical theory known as trigonometry purports to specify the length of the third side of a triangle once the lengths of the other sides included in the angle are given. The problem of measuring the two sides and the angle of a real physical triangle (say a tract of land) is not the business of the trigonometer. This is the business of the surveyor, who works with appropriate instruments. It so happens that small errors in the surveyor's measurements are reflected in a small error of the theoretically inferred result. This circumstance makes trigonometry useful as a practical science. But whether this is so or not, a theoretical science has its own validity: its conclusions are certain consequences of what has been given; they need not correspond to what is factually true.

Euclid's geometry is a more abstract science than the applied trigonometry used by the surveyor. A theorem of geometry says that "there exists" only one triangle with the lengths of two of its sides and the included angle specified. (This assertion is to be understood in the sense that all such triangles are congruent.) Elementary geometry does not specify how the length of the third side is to be determined from the lengths of the other two and from the included angle. It says only that the length of the third side *is* determined once the other two lengths and the angle are given. This conclusion is not "practical" in the business of surveying, because what the surveyor wants to know is how actually to find the length of the third side. Nevertheless the "existence" theorem of geometry is important, for only if the theorem is true does it make sense for the trigonometer to calculate the third side.

Game theoretical conclusions, like all mathematical conclusions, are based on givens and on the assumption that these givens can be somehow made known. The given payoffs are assumed to reflect the psychological worth of the associated outcomes to the player in question. The task of determining these psychologically meaningful payoffs is the task of the psychologist, not of the game theoretician, just as the job of measuring the distances along the surface of the earth is the job of the surveyor, not of the mathematician. It goes without saying that it helps the surveyor to be versed in mathematics. Similarly it may help the psychologist to be versed in game theory.

While the actual determination of the payoffs is not the game theoretician's concern, the game theoretician has something to say about the *scale* on which such determination should be made. We must therefore look into this question.

Scales of Measurement

All measurements specify relations among magnitudes. Once these relations have been specified, other relations can be derived from them. For example, if A is found to be twice B and B three times C, then we can conclude that A is six times C. Some relations are not so exactly specified, and the conclusions that can be derived from them are accordingly less precise. For instance, if all we know is that A is larger than B and B is larger than C, we can conclude that A is larger than C but not how much larger.

It stands to reason that the more exactly relations are specified, the more specific are the conclusions that can be drawn from them. But not all relations can be specified as exactly as we wish. For instance, if we read the temperature to be 20° F. on Monday and 40° F. on Tuesday, we can certainly say that it was warmer on Tuesday than on Monday. But we cannot say it was twice as warm on Tuesday. For suppose the temperatures had been read on a Centigrade scale. Then the readings would have been —6 2/3° C. and 4 4/9° C. for Monday and Tuesday, respectively. The second of these numbers is certainly not twice the first. Hence the statement "It was twice as warm on Tuesday as on Monday" is true only with respect to the Fahrenheit scale. But the Fahrenheit scale is not something given by nature; it was invented by Fahrenheit, who chose the fixed points and the size of a degree quite arbitrarily. A conclusion which depends on these specific conventions cannot reflect an objective state of affairs, such as the weather reflects.

It follows that it makes no sense to speak of the ratio of two temperatures, at least on the conventional scales. It is, however, meaningful to speak of the ratios of *differences* of temperature. Suppose, for example, we read 30° F. on Wednesday. Then we could say, "The rise of temperature was twice as much from Monday to Tuesday as the fall from Tuesday to Wednesday." Indeed on the Centigrade scale the temperature would have read −1 1/9° on Wednesday, and on that scale the rise in temperature from Monday to Tuesday was 11 1/9°; the fall from Tuesday to Wednesday was 5 5/9°. The first number is indeed twice the second.

If magnitudes are given in such a way that one can sensibly say of any two which is the larger but can say nothing about the magnitude of the differences, these magnitudes are said to be given on an *ordinal scale.*

Thus if I can say that I prefer Keats to Shelley and Shelley to Wordsworth (and therefore Keats to Wordsworth), then my preferences among poets are defined on an ordinal scale. But I may not be able to say whether my preference of Keats to Shelley is larger or smaller than my preference of Shelley to Wordsworth. If I cannot, then my scale of preference is no *stronger* than the ordinal scale.

If differences of magnitude can be compared, but nothing can be said about the ratios of the magnitudes themselves, the magnitudes are said to be given on an *interval scale*. For instance, the conventional temperature scales are interval scales, as we have seen.

If ratios of the magnitudes themselves can be specified, the magnitudes are said to be given on a *ratio scale*. Weight, money, length, area, energy, time intervals are all measured on ratio scales.

Of the three scales just described, the ordinal scale is the "weakest," and the ratio scale is the "strongest." The ordinal scale is weak in the sense that little information is given about the magnitudes measured on it—only the rank order of the magnitudes is given. This means that we have much freedom in assigning actual numbers to the magnitudes. Suppose, for example, the relative magnitudes of A, B, and C are given by the inequality $A > B > C$. We can, if we wish, assign the numbers 100, 17, and -25 to A, B, and C respectively. Or we can equally well assign the numbers 3, 2, and 21/11. It does not matter how we assign the numbers as long as they are in descending order; for this is the only relation specified by the ordinal scale of magnitudes. Mathematically speaking, one can substitute for A, B, and C any three other numbers A′, B′, C′, so long as the latter are obtained from the former by a so-called monotone transformation—a formula that changes the numbers of one set to the numbers of another set while preserving their relative positions on the axis of real numbers. Examples of such formulas, or mathematical functions as they are called, are $y = x^3$ or $y = 10x$. They are called positive monotone because the two variables always increase or decrease together. This property guarantees the preservation of the rank order. For example, if the x's are 100, 17, and −25, and if we transform the x's into y's by $y = x^3$, we obtain instead of 100, 17, and −25, 1,000,000, 4913, and −15,625 which are still in the same rank order. If we transform by $y = 10x$, we obtain 1,000, 170, and −250 which are again in the same rank order. But clearly the transformation $y = x^2$ will not do, because this gives 10,000, 289 and 625 which are not in the same order of magnitude as 100, 17, and −25. The function $y = x^2$ is not a monotone function of x. Mathematically speaking, the ordinal scale is *invariant with respect to positive monotone transformations.*

Consider now the ratio scale, which, we said, was the strongest of the three. If magnitudes are given on a ratio scale, this means that the ratios of any two magnitudes are also given. We now ask what kind of transformations leave the ratios the same. The answer is that only the function $y = ax$ where a is a constant leaves the ratios the same. If we demand that also the relative magnitudes are preserved, then a must be a positive constant. For example, if we multiply our three numbers 100, 17, and −25 by 10, we get 1,000, 170, and −250. The ratio between any two of them is clearly the same as before, because the factor of ten cancels out in the ratio. The transformation $y = ax$ is sometimes called the similarity transformation. The ratio scale is *invariant with respect to the similarity transformation.*

Finally, let us look at the interval scale. The important relation now is the ratio of differences. This relation is left invariant with respect to the so-called linear transformation $y = ax + b$. If the rank order is to be preserved, we must also have $a > 0$. The ratio of the differences among our three numbers is $(100 - 17)/[17 -$

$(-25)] = 83/42$. Let $a = 2$ and $b = 1$. Then the three numbers become 201, 35, -49. The ratio of the differences is now $(201 - 35)/[35 - (-49)] = 166/84 = 83/42$. Thus the ratio of differences has been preserved. The interval scale is *invariant with respect to the (order-preserving) linear transformations* $y = ax + b$ $(a > 0)$. (In what follows always assume that a linear transformation is order-preserving.)

In game theory it is assumed that magnitudes assigned to the payoffs denote the worth of the payoffs to the respective players, and denote also that the magnitudes can be determined on an interval scale. This means that a player given any three outcomes, A, B, and C, can say without ambivalence not only his order of preference (say, $A > B > C$) but can also tell the ratio of the differences among the preferences, for example, "My preference of B over C is $3\frac{1}{2}$ times greater than my preference of A over B."

Thus the player's preference scale used in game theory must be stronger than the ordinal scale (on which he needs to specify only the preference order among the outcomes) but need not be as strong as the ratio scale (on which he would have to specify the ratio of any two magnitudes).

Payoffs as Utilities

When payoffs are specified on an interval scale, they are called utilities. Let us now see why it is necessary to specify payoffs on the interval scale.

Suppose a certain game of strategy is analyzed to the extent that a best way of playing the game is found. Suppose now the game is played for different stakes. What had been pennies become dollars. It is quite conceivable that real people would play the same game quite differently if it were played for dollars than if it were played for pennies. But how differently they would play we do not know without knowing quite a bit about the psychology of playing for money and, very possibly, without knowing a great deal about different kinds of people. Some people might take smaller risks when playing for bigger stakes; some might take larger risks, carried away by the excitement of playing for big stakes. Game theory, however, is not at all concerned with such matters. It is concerned with *rational* ways of playing and this means only one thing: to get as much as possible in terms of utilities. Therefore it should make no difference in what units the utilities are measured. Translated into mathematical language, this means that the payoff scale should remain invariant under the positive similarity transformation $y = ax$ $(a > 0)$.

Next, suppose all the possible payoffs of a player are changed by having the same constant added to each of them. This means, essentially, that whatever the outcome of the game, the player gets a certain fixed bonus (or pays a certain fee). Since the magnitude of the bonus or of the fee is independent of the outcome, including the reward or the penalty in the payoff should make no difference in the way the game should be rationally played by the player in question. Therefore the payoff scale should also be invariant with respect to the transformation $y = x + b$ (b constant). Combining this transformation with the similarity transformation, we get a linear transformation in the form $y = ax + b$. But if a scale remains invariant under the linear transformation, it is an interval scale. This is why utilities are assumed to be given on an interval scale, namely to make them invariant under linear transformations.

How realistic is this assumption? Recall that it implies the player's ability to measure the ratios of differences among his preferences. If the nature of the outcomes is not restricted, this may mean that the player must be able to answer questions like "How many times greater is your preference for malaga over madeira than your preference of madeira over port?" To expect meaningful (i.e., consistent) answers to questions of this sort does not seem realistic.

Actually the measurement of utilities is supposed to be carried out in another way. Suppose the payoffs are a bottle of malaga, a bottle of madeira, and a bottle of port, and suppose a player prefers wines in that order. This determines his preferences on an ordinal scale, but we need a stronger one, namely an interval scale. To obtain it, we offer the player a choice between two tickets. One ticket entitles him to a bottle of madeira. The other ticket is essentially a lottery ticket. It will get him a bottle of malaga or a bottle of port depending on the outcome of an event determined by chance, where the chances are, say, 50-50 of either outcome. Now all the man has to decide is which ticket he prefers. No *numerical* estimate is demanded of him.

Suppose, for example, the man prefers the lottery ticket to the ticket which entitles him to madeira. Let him now be confronted with another choice, the same as before except that the chances in favor of malaga (his favorite wine) have now been reduced to forty percent. If he still prefers the lottery ticket, let the chances for malaga be reduced still further. Somewhere along the line before the chances for malaga are reduced to zero, the man must change his mind; otherwise he will be preferring a ticket which entitles him to port over a ticket which entitles him to madeira, in contradiction to his declared preference.

Suppose the change of mind occurs just as the chances for malaga reach twenty-five percent. The game theorist now determines the man's utilities (i.e., his preferences on an interval scale) as follows. Let u_0 stand for the utility of port, u for the utility of madeira, and u_1 for the utility of malaga. We can, if we wish, assign the value 0 to u_0 and the value of 1 to u_1, because on an interval scale the choice of the zero point and the unit point are arbitrary. It remains merely to determine the value of u, or, which is the same thing, the ratio of the intervals $(u_1 - u)$ and $(u - u_0)$. This is done by equating the *expected*[3] utility associated with lottery ticket to the utility of madeira, namely,

$$.25u_1 + .75u_0 = u. \qquad (1)$$

But we have set $u_1 = 1$, $u_0 = 0$. Therefore $u = .25$. On the interval scale of utilities, then, the preference magnitudes of malaga, madeira, and port are respectively 1, .25, and 0. Since the interval scale is invariant with respect to linear transformations, any other three numbers will do provided they are obtained from 0, .25, and 1 by a linear transformation, for example, 3, 4, and 7 or 13, 17, 29: in short, any three numbers such that the difference of the larger two is three times the difference of the smaller two. For by indicating his indifference between the (.25, .75) lottery ticket and madeira, the man has effectively said "My preference of malaga over madeira is three times greater than my preference of madeira over port." Possibly he could not have given such an answer directly.

Now whether in fact utility scales can be teased out by such methods more easily than by direct questions about relative sizes of preference differences (if at all) is an open question. The answer depends on whether the ratios so determined are consistent. Suppose an interval scale is obtained on which several objects are rated, A, B, C, D, etc. Then, if our investigation reveals that $(A\text{-}B) = 2(B\text{-}C)$ and $(B\text{-}C) = 3(C\text{-}D)$, then we ought to obtain $2(A\text{-}B) + 2(B\text{-}C) = 2(A\text{-}C) = 6(B\text{-}C) = 18(C\text{-}D)$ or $(A\text{-}C) = 9(C\text{-}D)$. But the last relation can be determined also directly by using A, C, and D as the triple (offering a choice between C with certainty and various lottery tickets involving A and D). The two results ought to be consistent with each other if the utilities of the four objects are to be meaningful. It is by no means certain that data obtained from people asked to express preferences of this sort will be sufficiently consistent to allow the construction of an unambiguous utility scale.

As has already been pointed out, these matters are not of concern to the game theoretician. His position is that *if* utility scales can be determined, *then* a theory of games can be built on a reliable foundation. If no such utility scale can be established with references to any real subjects, then game theory will not be relevant to the behavior of people in either a normative or descriptive sense.[4]

For the time being, we shall leave this question of relevance and assume that somewhere players exist or can be imagined for whom utility scales of payoffs can be constructed.

Choices Between Lottery Tickets

The utility scale, as it has been defined, implies that of two *risky* outcomes, the one with the greater expected utility is always preferred. A risky outcome can be thought of essentially as a lottery ticket entitling the owner to any of several prizes depending on the outcome of a chance event. Such a lottery ticket carries a list of the prizes with the probability attached to each, namely the probability of the event which entitles the holder of the ticket to the prize in question. Two examples of such lottery tickets are given:

	PRIZES	PROBABILITIES (chances to win)
Lottery Ticket #1	Prepaid trip to Mexico City	(.02)
	Ford station wagon	(.10)
	Getting elected mayor	(.00)
	20 days in jail	(.20)
	Nothing	(.38)
	Win $1,000	(.19)
	Lose $1,500	(.11)

Lottery Ticket #2		
	Prepaid trip to Mexico City	(.08)
	Ford station wagon	(.00)
	Getting elected mayor	(.12)
	20 days in jail	(.35)
	Nothing	(.10)
	Win $1,000	(.15)
	Lose $1,500	(.20)

The definition of utility, as it is used in game theory, implies that players can consistently choose among all possible lottery tickets of this type.[5] Let us take a closer look at what this means.

Note that the outcomes listed on one ticket are all listed also on the other. This can always be done if we include outcomes with zero probabilities. For example, "getting elected mayor" has zero probability on Ticket #1, and "Ford station wagon" has zero probability on Ticket #2. We could, therefore, have omitted "getting elected mayor" from one ticket and the station wagon from the other. But this also implies that we can add to any lottery ticket any item which is not listed on it, if we assign zero probability to the added item. This, in turn, means that the lists of prizes associated with all the relevant lottery tickets can be made identical.

Imagine now a man who must choose between Tickets #1 and #2. Suppose Mexico City, station wagon, getting elected mayor, and winning $1,000 are attractive to this man, while 20 days in jail, losing $1,500, and nothing are not attractive. Some of the attractive items appear with greater probabilities on one ticket, some on the other, and similarly for the unattractive items. What will determine the man's preference? He may concentrate on one item, for example, 20 days in jail. If he wants to avoid this at all costs, he will be wise to choose Ticket #1. But if he happens to want to get elected mayor at all costs, he will choose Ticket #2. This sort of "one-criterion decision" is easy, but it ignores the really important problems of conflicting preferences. In order to face these problems, the man must weigh advantages against disadvantages. If he wants very much to be elected mayor but also to stay out of jail, he will have to estimate the relative importance to him of these outcomes so as to know whether an increase of the probability of being elected mayor, from zero to .12, is worth the increase of the probability of jail from .20 to .35. This estimate involves comparisons of two pairs of outcomes but of course neglects all the others. We begin to see how complex the problem becomes if all of the outcomes must be considered.

Now, in being asked to decide between the two tickets, the man is not asked to calculate anything. He is asked simply to choose between the two. It is from his *choices* that the theoretician will construct the man's utility scale on which all the outcomes will be assigned numbers (utilities). But this scale can be constructed only if the man's choices are consistent, e.g., if having chosen Ticket #2 over Ticket #1 and Ticket #3 (not shown here) over Ticket #2, the man will choose Ticket #3 over Ticket #1 when this pair is presented to him for choice. This consistency is not expected in experiments or in real life situations. Faced with complex choices, people are very frequently inconsistent, especially when the outcomes are as different qualitatively as the ones we have listed (and frequently also when the outcomes are easily comparable, e.g., all money prizes with different probabilities attached). Nevertheless, game theory postulates "rational players" who are able to choose consistently among all possible risky outcomes. The question whether such players exist bears on the question of whether game theory can ever be applied to behavior (either normatively or descriptively). However, many a theory deserves examination quite apart from the question of its immediate applicability, and game theory is an outstanding example of a theory of this sort.

Expected Gain

Consistent choices among risky outcomes become somewhat more likely if the outcomes are all of one kind. Suppose all outcomes are in money, and the man must choose between the following lottery tickets.

	PRIZES	PROBABILITIES
Ticket A	Lose \$10	(.05)
	Lose \$5	(.10)
	Lose \$1	(.20)
	Nothing	(.35)
	Win \$2	(.25)
	Win \$20	(.03)
	Win \$100	(.02)
Ticket B	Lose \$10	(.15)
	Lose \$5	(.20)
	Lose \$1	(.20)
	Nothing	(.30)
	Win \$2	(.00)
	Win \$20	(.08)
	Win \$100	(.07)

Here a man might calculate his "expected gain" from each lottery. To do this he multiplies each gain (positive or negative) by the associated probability and adds the products. Thus the expected gain associated with Ticket A in dollars is

$$(-10)(.05) + (-5)(.10) + (-1)(.20) + (2)(.25) + (20)(.03) + (100)(.02) = \$1.90$$

The expected gain associated with Ticket B is

$$(-10)(.15) + (-5)(.20) + (-1)(.20) + (20)(.08) + (100)(.07) = \$5.90$$

The man may well choose Ticket B on that basis. Clearly the choice is rational if it is offered many times. For then it is highly probable that the expected gain of a ticket will be the actual gain averaged over many such tickets. In fact, the maximization of expected gain is actually used as a decision principle in businesses where actuarial computations are made, e.g., insurance.

However, if the choice between Ticket A and Ticket B is offered only once, it is by no means certain that everyone will prefer Ticket B, nor that it is necessarily the rational choice. For on the single occasion neither of the expected gains will be realized. What will be realized is one of the listed outcomes. In this case the expected gain is nothing but a theoretical construct without practical significance: it is no consolation to know that the expected gain was $5.90 if you actually lose $10, which is more likely if Ticket B is chosen.

A man faced with a choice between Ticket A and Ticket B might be impressed by the fact that Ticket A offers him thirty chances out of a hundred to win money, while Ticket B offers only fifteen chances out of a hundred. But if he is guided by this consideration, he is in effect simply making a cruder calculation than that based on expected gain. Specifically, he is equating the utilities of all the winning outcomes. Is this rational? The theory of utility, as it is presented in game theory, makes no judgment about such matters. How one assigns utilities to outcomes is the decision-maker's private affair. These assignments are to be presumably *discovered* from the decision-maker's preferences among all possible lottery tickets, provided these preferences are consistent. Therefore it makes no sense to ask whether it is rational always to choose risky outcomes with greater expected *utility*. Utility, as it is defined in game theory, is that quantity whose expected gain in a risky choice is attempted to be maximized by a decision-maker, whose choices among risky outcomes are consistent. In short, in the context of game theory it makes no more sense to ask whether maximization of expected utility gain is a rational principle of decision than to ask whether an inch is really an inch long.

This view of utility is not always well understood, and so must be made entirely clear before game theory proper is examined. As an example, suppose a man is asked to bet his life savings ($10,000) on a toss of a die, which will win him $100,000 if the ace turns up. The expected gain (in money) of this bet is $8,333.33. Most people would feel that it would not be wise to accept this bet. This feeling is a reflection of a tacit assumption that the loss of $10,000 in life savings has a far greater disutility than one-tenth of the utility of winning $100,000. On the other hand, suppose the man needs $100,000 to ransom his child from murderous kidnappers and that this bet is the only way he can procure the amount. His acceptance of the bet becomes understandable.

Another important problem raised in connection with the use of utilities is that of comparing the utilities for the same outcome of different people.

Suppose two people, A and B, must decide jointly between two alternatives, say between two business ventures X and Y. Should they undertake venture X, A will make $200 while B will make $400. If they undertake venture Y, A will make $500, while B will make $200. It stands to reason that A prefers Y, while B prefers X. One might argue that it makes more sense for them to undertake Y, since *jointly* they will make more in that venture. It would, indeed, be worthwhile to A to pay B $250 in order to induce him to join in venture Y, since as a result A will have made $50 more than in venture X, and so will B. This arrangement is reasonable if (1) it is possible for the money to be transferred, (2) agreements about the transfer are enforceable, and (3) in being transferred, money preserves its utility. Situations can be easily imagined where any or all of these conditions are violated. Moreover, outcomes of decisions are not always associated with monetary payoffs.

Suppose, for example, venture Y will bring A a much larger profit but will cost B his self-respect. It is now more difficult to argue that A can compensate B by turning over a portion of his profits to him. To be sure, there are versions of economic theory in which all utilities can be reduced to a common measure like money; but the realism of such theory is open to question. At any rate it would be desirable to have a theory of rational conflict in which such an assumption need not be made.

Now if the utility of each party (a party is represented by a set of interests) is given on an interval scale, the question of comparing the utilities for a given outcome of two or more parties need not arise, in fact, cannot arise, for in that case such comparisons are meaningless.[6]

Whether the assumption of an interval scale for utilities is psychologically realistic is another question, and one against which much evidence can probably be marshalled. This assumption is made in order to avoid psychological issues, such as whether utilities of different parties can be meaningfully compared.

Having by-passed all the psychological problems associated with the measurement of utilities or, indeed, with the question of whether consistent utility scales can be established at all, two-person game theory makes only one demand, namely that whatever utility scale is established, it shall be an *interval scale*. This means that whatever decisions are prescribed in risky choices, they should remain the same if the decision-maker's utility scale undergoes a linear transformation. If the utilities associated with a certain set of outcomes are denoted by $u(o_i)$, where $o_1, o_2, \ldots o_n$ are the outcomes, no conclusion of the theory should be affected if instead of $u(o_i)$ we write $au(o_i) + b$ where a and b are arbitrary constants, $a > 0$. Moreover, we can choose a and b differently for each of the players in a game.

We shall henceforth accept this basic assumption of two-person game theory.

3. Strategy

Race-to-Twenty is a children's game which is a simple version of the ancient game of Nim. In our discussion the extreme simplicity of Race-to-Twenty will enable us to analyze the game completely. The idea of strategy will come out of this analysis.

In Race-to-Twenty, two players count from 1 to 20 alternately, each player having a choice of advancing the count by either one or two numbers. Thus the first player may say "One" or "One, two." If he says "One," the second player may say "Two" or "Two, three"; if the first player says "One, two," the second player may say "Three," or "Three, four," etc. The player who says "Twenty" first wins.

A little reflection shows that player 1 must always win. He will certainly win if he can say "Seventeen." For in that case, if player 2 says "Eighteen," the first player can say "Nineteen, twenty"; if the second player says "Eighteen, nineteen," the first player can say "Twenty." Similarly the first player can always say "Seventeen" if he can say "Fourteen"; he can say "Fourteen" if he can say "Eleven," and so on with "Eight," "Five," and "Two."

But player 1 can start the count with "One, two." Therefore, he will always win, provided he ends his successive counts on five, eight, eleven, etc. To win, he must adhere to the following *policy:* always end the count with a number which upon division by 3 gives remainder 2. The critical numbers, 2, 5, 8, 11, 14, 17, and 20 all give remainder 2 upon division by 3. The mathematician says these numbers are *congruent to 2 modulo 3.* Two numbers congruent to a third are congruent to each other. We can, therefore, also formulate the policy so: always end the count on a number congruent to 20 modulo 3.

This idea of "congruence modulo some number" immediately enables us to generalize the winning policy of any game of this sort, where the target number may be chosen at will and each player has a choice of advancing the count by any of a set of successive numbers. Suppose, for example, the target number is 57, and each player can advance the count by one, two, three, four, or five consecutive numbers. Then the winning policy for the first player is the following: always end the count by a number congruent to 57 modulo 6. And in general, if the target number is N and the count may be advanced by 1, 2, . . . or k, the first player will win if he ends all his counts on numbers congruent to N modulo k + 1.

The statement, always end the count on a number which . . . etc., is clearly a set of directions which tells a player what he is to do *in every possible situation* in which he may find himself while playing one of the games just described. A strategy is essentially just that. It is, of course, not always possible to give such instructions in so compact a form. For example, in order to state a strategy for the game of Tic-Tac-Toe, one needs many more words than in the case of Race-to-Twenty, because we do not have terms to express the various classes of configurations which may arise. Specifically, the description of a strategy chosen by player 1 (Naughts) might start as follows. "My first naught will be placed in the center square. Should Crosses (player 2) play in a corner, I will play in the diagonally opposite corner. . . ." This does not complete the description, which becomes more and more involved as it proceeds, because the next choice must be given in all the possible situations which may arise. These become progressively more numerous, because Naughts must take into account *all possible* choices of Crosses, stupid ones as well as intelligent ones. After this is done, *one* strategy has been described.

How many of such strategies are there in a game? How many would there be in a game like Tic-Tac-Toe, and in how many different ways can a game of Tic-Tac-Toe be played? Although the latter number is quite large, it does not compare in magnitude with the number of possible strategies. This is important to keep in mind, because it is easy to confuse the idea of strategy with that of a particular realization of a game, which is determined by a *pair* of strategies.

In Tic-Tac-Toe, the first player has a choice of nine boxes in which to put his naught. Then the second player has a choice of eight boxes in which to put his cross. Then the first player has a choice of seven boxes in which to put his naught, etc. The game ends, of course, when one of the players has placed three of his marks in a row. This can happen at the earliest on the fifth move. Therefore the game can last anywhere from five to nine moves. Actually a game of Tic-Tac-Toe is usually broken off when it becomes obvious that one of the players must win or that neither can win. But if we are to go by this criterion, then every game of Tic-Tac-Toe should be broken off before it starts, since a complete analysis reveals that neither player can win if both play correctly. We shall suppose, however, that every game is played out to the bitter end, and for simplicity of calculation we shall suppose that the game continues until all the boxes have been filled, even if a player has succeeded in getting three in a row. So calculated, the number of different ways a game of Tic-Tac-Toe can be played is $9 \times 8 \times 7 \times 6 \times 5 \times 4 \times 3 \times 2 = 362{,}880$ ways. This number is inflated, because it includes all the plays of the game that have continued even after one player has won. The number is inflated also in another way. It includes plays of the game which show the same *record* (i.e., plays where naughts and crosses are in the same boxes but which happen to have been arrived at by different sequence of choices). We can easily correct for this by calculating the number of ways five naughts can be put into nine boxes (for this leaves the other four boxes for crosses). This number turns out to be 126, much smaller than our original estimate of 362,880. But even 126 is an inflated value if one does not wish to distinguish between outcomes which can be considered equivalent. For example, compare the two outcomes shown in Figure 1.

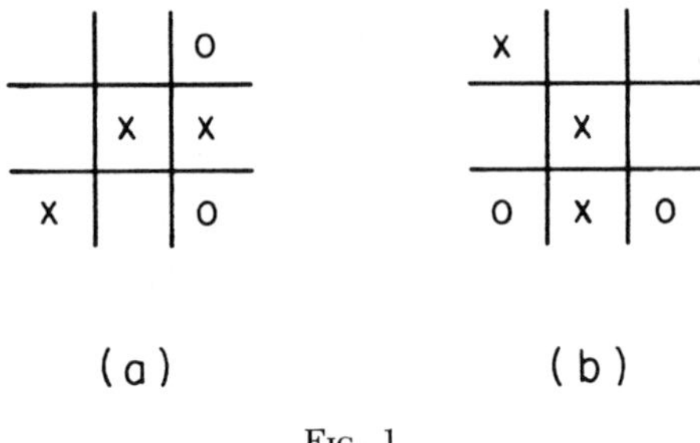

FIG. 1

Obviously (b) can be obtained from (a) by rotating (a) clockwise through 90°. Should not therefore (a) and (b) be considered the same outcome? This depends, of course, on the purpose of the classification, i.e., the sort of analysis the classification is supposed to serve.

We see, then, that the question of how many different ways a game can be played does not have a clearly defined answer unless we specify how we wish to classify the different outcomes.

With strategies the situation is similar. In order to fix ideas, we shall calculate the largest possible value of the number of strategies possible in Tic-Tac-Toe, for this number is the easiest to calculate.

We note that the first player can start the game in nine different ways. On the next move, the second player has eight choices, after which the first player has seven choices. Now the number of ways in which the first player can specify what he will do on the third move in response to each of the other's choices is 7^8. Similarly the number of ways in which the first player can specify what he will do on the fifth move in response to the other's six choices is 5^6, and the number of ways in which the first player can specify what he will do on the seventh move in response to each of the other's four choices on the sixth move is 3^4. On the ninth move, however, the first player can do only one thing, namely put a naught in the only box left. Thus the number which is sufficient to encompass all of the first player's strategies is $9 \times 7^8 \times 5^6 \times 3^4 = 65{,}664{,}686{,}390{,}625$.

Like our previous estimates of the number of outcomes, this number is strongly inflated. In order to trim it down, we would have to take into account the symmetries inherent in the structure of the game. This is by no means an easy task. Therefore, the question, "How many different strategies are available to each of the players in Tic-Tac-Toe?" remains open. The answer depends on which strategies we are willing to consider equivalent, for practical purposes. "For practical purposes" is to be understood in terms of what is important in the strategic structure of the game.

The importance of the idea of strategy stems not from the possibility of analyzing a game but rather from the resulting *conceptualization* of a game. In common parlance we speak of flexible and of rigid plans of action. We call a plan of action rigid if it contains few contingent decisions, and flexible if it contains many such decisions. Plans made long in advance of the action we associate with rigidity, implying that to be flexible, one must be willing to wait and see, to defer decisions until one has taken into account the way a situation develops. This conception of flexibility may be valid in real life, but it is irrelevant to the notion of strategy, as the term is defined in game theory. One might think that making a commitment to a definite strategy before the play of the game starts is tantamount to abandoning flexibility, but this is by no means the case. A strategy, as we have defined it here, *already* contains in it all the contingencies which *can possibly* arise. Deferring decisions until the corresponding choice must be made does not increase flexibility. Any feeling to the contrary which we may have stems from the well-known circumstance that in real life *unforeseen* developments occur. In the situations defined as games in game theory there are no such unforeseen developments. Everything that can *possibly* occur in the course of a play of a game is known. What is not known is the way the opposing player will *actually* choose each of his moves. But all the *choices* open to him are known. A strategy is a plan which provides for every possible choice on the part of the other player.

To emphasize the irrelevance of the notion of rigidity to that of strategy choice, consider the following strategy chosen by player 1 in Race-to-Twenty. "Whatever he does, I shall continue the count by two." Our immediate reaction to this strategy is that it is too rigid. We feel the player should make his responses contingent on the other player's responses. For example, if the other says "Ten" it is better to continue with one number and say "Eleven." And if he says "Nine," it is better to continue with two numbers and say "Ten, eleven." However, this is merely one way of describing a winning strategy. Another way of describing it follows: "Whatever the other does, I shall end my counts on numbers congruent to 2 modulo 3." Actually this strategy seems no more flexible than the preceding one, but it nevertheless encompasses a maximum of flexibility, because it has made provision for every possible eventuality.

In short, what we mean by flexibility in common parlance is really bound up with the inherent uncertainty of the real world. Things are always coming up *not* listed in the contingencies with which we are familiar. Flexibility means remaining uncommitted to the extent of allowing oneself some leeway to design ways of dealing with these unforeseen events. Such events are deliberately excluded from consideration in game theory. This, as well as the many other simplifications assumed in game theory, may have an important bearing on the extent to which game theory can be applied to practical affairs, but no bearing on the conceptual value of game theory.

However large the number of strategies, they are not infinite, so long as the number of possible moves and the number of choices on each move remains finite. Next, it should be clear that once each player has chosen a strategy among all the strategies available to him, the course of the game is completely determined. This is not to say that each player is in a position to know at the beginning of the game how it will turn out. He cannot know this without knowing the other player's strategy as well as his own. But an observer who knows which *pair* of strategies has been chosen (one by each player) can, in principle, figure out the exact course of the game. The qualification "in principle" is, of course, crucial, because the vast numbers of strategies and the practical impossibility of specifying even a single strategy in words (for any but the most trivial games) makes it humanly impossible to carry out the necessary calculation. It cannot be emphasized too strongly, however, that detailed analysis of specific games is not a test of game theory. The theory is concerned only with the *logic* of strategic analysis. Accordingly, the fact that a single choice of a strategy by each player is in principle sufficient to determine the course of the play, and hence the outcome, enables the game theoretician to continue the analysis *on another level.* Instead of worrying what is likely to happen on each move, while the possibilities become super-astronomically numerous, the game theoretician can reduce the question to what will happen as a result of a choice by each player of a single strategy.

The number of strategies available to each player, as we have seen, is finite. Suppose, therefore, the strategies available to player 1 are numbered from 1 to N, and those open to player 2 from 1 to M. Each pair of strategies chosen can be designated by (i, j), meaning that the i-th strategy was chosen by player 1 and the j-th by player 2. This pair determines an outcome O_{ij}. The mathematician says that the variable outcome O_{ij} is a function of two variables, i and j. If we arrange all the possible outcomes O_{ij} (some of which may well be identical) in a rectangular array of N rows ($i = 1, 2 \ldots N$) and M columns ($j = 1, 2, \ldots M$), we shall have the corresponding game represented by a *matrix,* which depicts the entire strategic structure of the game. When games are classified and conclusions about them are drawn according to the properties of these matrices, the resulting theory is called the *theory of games in normal form.* Practically the entire theory of two-person zero-sum games is stated as a theory of games in normal form. The next four chapters will be devoted to this topic. The paradigm of a game in normal form is shown as Game 1.

	1	2	....	j	....	M
1	O_{11}	O_{12}	. . .	O_{1j}	. . .	O_{1M}
2	O_{21}					
.						
i	O_{i1}	. . .	. . .	O_{ij}		
.						
N	O_{N1}	. . .	. . .	. . .	. . .	O_{NM}

Game 1

4. *The Game Tree and the Game Matrix*

One of the keystone ideas of game theory is that any "well defined" game can be reduced to normal form. As we have said, a game in normal form presents to each player a choice among several alternatives called *strategies*. When such a choice is presented, it suffices for each player to choose one of the strategies available to him. It is essential that every player make his choice simultaneously or, what is the same thing, *in ignorance* of what choices are made by others. (We shall examine below the consequences of violating this condition.) The simultaneous choices of respective strategies determine an outcome of the game, which, in turn, specifies the payoff to each player. These payoffs may be positive or negative and are supposed to be represented in units of utility.

Typical parlor games are, of course, not described in normal form. Instead they are defined by *rules*, among which the most essential are those that specify what choices a player has in each possible "position" resulting in the *course* of the game. So conceived, a play of a game is a sequence of moves or choice points, at which now one, now another player must choose among available alternatives, i.e., the legal moves. This form of the game also has a representation called the *extensive form* of the game. We can now restate the first sentence of this chapter as follows: every game can be reduced from the extensive form to the normal form. This implies what is meant by a "well defined" game, namely a game which can be represented in extensive form.

Let us see how the extensive form is obtained and how the reduction to normal form is made. We cannot take any "real" game as a working example, because, as we have seen, even a very simple child's play like Tic-Tac-Toe would involve an enormous number of strategies if put into normal form. We must therefore confine our examples to artificial games, invented specifically for illustrative purposes. Although such games must necessarily be much simpler than even Tic-Tac-Toe, they illustrate the basic principles equally well.

We shall analyze a game called Square-the-Diagonal. It is played by two players as follows. The first player, whom we shall call Castor, has a choice of 1, 2, or 3 units. When he has made his choice, the second player, Pollux (knowing what the first player has chosen), has a choice of 1 or 2 units. Following this, Castor, knowing how Pollux has chosen, again has a choice of 1, 2, or 3 units. Imagine that each choice determines respectively the length, the width, and the depth of a box. The box determined by a play of the game will have dimensions x, y, and z, corresponding to the three consecutive choices of the two players. The square of the length of its diagonal, therefore, will be $x^2 + y^2 + z^2$.[7] The payoffs to each of the players will be determined by what this magnitude turns out to be. Namely, if for a particular outcome $x^2 + y^2 + z^2$ is congruent to 0 or 1 modulo 4 (i.e., leaves a remainder of either 0 or 1 upon being divided by 4), then the amount $x^2 + y^2 + z^2$ will be won by Castor. If

$x^2 + y^2 + z^2$ is congruent to 2 or 3 modulo 4, that amount will be won by Pollux.

Our game is now completely defined and can be represented in extensive form. This is done by constructing the *game tree*, a diagram in which all the choices and the resulting positions, including the final outcomes, are displayed. Square-the-Diagonal is shown in extensive form in Figure 2.

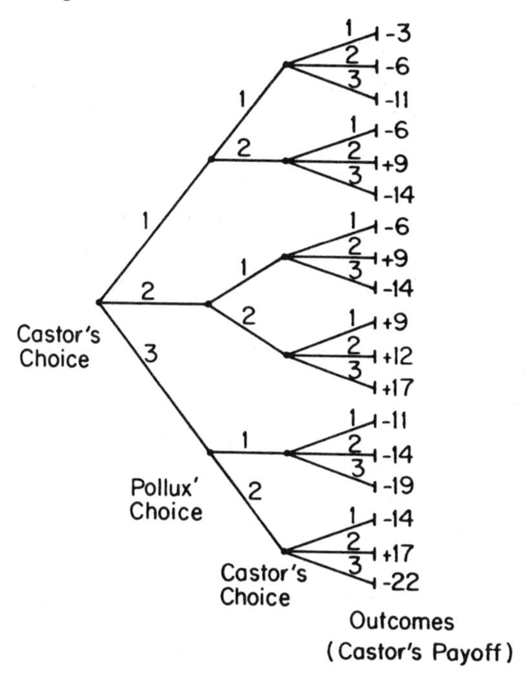

Fig. 2. The Square-the-Diagonal game in extensive form.

From the game tree we see that if Castor starts the game by choosing 1 unit, Pollux can always be sure of winning something, for Pollux can choose 1 unit, which gives Castor a choice among losing 3, losing 6, or losing 11, depending on whether he chooses 1, 2, or 3 units for the depth of the box.

Similarly, if Castor starts with 3 units, Pollux can again make sure that Castor will lose something, namely by choosing 1 unit, which gives Castor a choice of losing 11, 14, or 19.

The matter looks different if Castor starts by choosing 2 units. For then, if Pollux continues with 1 unit, Castor can win 9 (by choosing 2 units on the final move); if Pollux continues with 2 units, Castor can win as much as 17.

It follows that Castor can *guarantee* himself a win of 9 by starting with 2 units. Pollux can do nothing to prevent Castor from winning at least 9. If Pollux is not careful, he may lose as much as 17. On the other hand, Castor cannot hope to get more than 9 unless Pollux makes a stupid choice. For if Castor starts with anything but 2 units (say, with 3 in the hope of getting 17), Pollux can win as much as 11 by continuing with 1 unit. To obtain 17, Castor must count on Pollux's cooperation, i.e., on his continuing with 2 units following Castor's 3. But there is no reason for Castor to hope that Pollux will do this.

So far our analysis has proceeded sequentially from move to move. Let us now reduce the game to normal form so that a play of the game can be represented by a single simultaneous choice of strategy by each of the players.

It turns out that Castor has twenty-seven strategies. The first nine of these are as follows.

C_1: Choose 1 unit on each of the first and third moves, regardless of how Pollux chooses on his move.

C_2: Start with 1. On the third move do what Pollux has done on the second.

C_3: Start with 1. On the third move do the opposite of what Pollux has done on the second.

C_4: Start with 1. On the third move choose 2, regardless of how Pollux has chosen.

C_5: Start with 1. On the third move respond with 2 to Pollux' 1 and with 3 to his 2.

C_6: Start with 1. On the third move respond with 3 to Pollux' 1 and with 2 to his 2.

C_7 Start with 1. On the third move choose 3 regardless of Pollux' choice.

C_8: Start with 1. On the third move respond with 1 to Pollux' 1 and with 3 to his 2.

C_9: Start with 1. On the third move respond with 3 to Pollux' 1 and with 1 to his 2.

The next nine strategies prescribe the same contingencies respectively as the first nine but they all prescribe 2 as Castor's first move. The last nine prescribe 3 as Castor's first move and thereafter follow the same pattern as all the others. This completes the twenty-seven strategies available to Castor.

Pollux has eight strategies. Each of them can be designated by three numbers, each number (1 or 2) being understood as the response which Pollux will make to Castor's choice of 1, 2, or 3 units respectively. Thus Pollux' eight strategies are the following:

P_1: 1, 1, 1
P_2: 1, 1, 2
P_3: 1, 2, 1
P_4: 2, 1, 1
P_5: 2, 2, 1
P_6: 2, 1, 2
P_7: 1, 2, 2
P_8: 2, 2, 2

Combining Castor's twenty-seven and Pollux' eight strategies into a matrix, we have the game of Square-the-Diagonal in normal form see (see Game 2). The payoffs are the entries in the boxes of the matrix. By convention, only the payoffs to Castor (the chooser of the strategies

	P_1	P_2	P_3	P_4	P_5	P_6	P_7	P_8
C_1	−3	−3	−3	−6	−6	−6	−3	−6
C_2	−3	−3	−3	9	9	9	−3	9
C_3	−6	−6	−6	−6	−6	−6	−6	−6
C_4	−6	−6	−6	9	9	9	−6	9
C_5	−6	−6	−6	−14	−14	−14	−6	−14
C_6	−11	−11	−11	9	9	9	−11	9
C_7	−11	−11	−11	−14	−14	−14	−11	−14
C_8	−3	−3	−3	−14	−14	−14	−3	−14
C_9	−11	−11	−11	−6	−6	−6	−11	−6
C_{10}	−6	−6	9	−6	9	−6	9	9
C_{11}	−6	−6	12	−6	12	−6	12	12
C_{12}	9	9	9	9	9	9	9	9
C_{13}	9	9	12	9	12	9	12	12
C_{14}	9	9	17	9	17	9	17	17
C_{15}	−14	−14	12	−14	12	−14	12	12
C_{16}	−14	−14	17	−14	17	−14	17	17
C_{17}	−6	−6	17	−6	17	−6	17	−6
C_{18}	−14	−14	9	−14	9	−14	9	−14
C_{19}	−11	−14	−11	−11	−11	−14	−14	−14
C_{20}	−11	17	−11	−11	−11	17	17	17
C_{21}	−14	−14	−14	−14	−14	−14	−14	−14
C_{22}	−14	17	−14	−14	−14	17	17	17
C_{23}	−14	−22	−14	−14	−14	−22	−22	−22
C_{24}	−19	17	−19	−19	−19	17	17	17
C_{25}	−19	−22	−19	−19	−19	−22	−22	−22
C_{26}	−11	−22	−11	−11	−11	−22	−22	−22
C_{27}	−19	−14	−19	−19	−19	−14	−14	−14

Game 2

shown as horizontal rows) are entered, it being understood in this case that the payoffs to Pollux are numerically the same as those to Castor but with the sign reversed.

In our analysis of the game in extensive form we have seen that Castor should start with 2. Thereafter, if Pollux continues with 1, as he should, Castor ought to follow with 2. He will then win 9, the amount which the game guarantees to him. If Pollux should (foolishly) play 2, following Castor's 2, then, of course, Castor should finish with 3 to get 17. This eventuality should also be foreseen in Castor's calculation. Therefore Castor's best strategy is the following: start with 2. If Pollux plays 1, play 2; if 2, play 3. This is designated as C_{14} in the game matrix.

What is Pollux' best strategy? If Castor should start with 1, obviously Pollux should reply with 1, thus insuring a positive payoff for himself (negative to Castor). If Castor should start with 2, Pollux should avoid 2, for this would give Castor the opportunity of winning 17. Pollux' best answer to Castor's 2 is 1. If Castor should start with 3, Pollux' best answer is again 1, for this too assures Pollux a positive payoff. Therefore Pollux' best strategy is the following: regardless how Castor starts, play 1. We have symbolized this strategy by the triple (1, 1, 1). It corresponds to strategy P_1 in the game matrix.

Consulting the game matrix, we see that the outcome determined by the players' simultaneous single choice of their respective best strategies is identical with the outcome predicted by the analysis of the game in extensive form. This is the outcome found at the intersection of Castor's strategy C_{14} and Pollux' strategy P_1.

5. *Dominating Strategy and Minimax*

Referring once more to Game 2, let us compare Castor's strategy C_{14} with C_{12}. We see that regardless of which strategy is chosen by Pollux, Castor will do at least as well with C_{14} and possibly better (if, for instance, Pollux should choose P_3) than with C_{12}. We say that C_{14} *dominates* C_{12}. Similarly, Pollux' strategy P_1 dominates P_3, since Pollux does at least as well with P_1 and possibly better (e.g., if Castor chooses C_{10} or C_{15}).

We said that P_1 is Pollux' best strategy. This does not mean, however, that P_1 dominates every other strategy available to Pollux. For example, P_1 does not dominate P_2. Pollux does not do better, or even as well, with P_1 *regardless* of Castor's choice. Should Castor choose C_{23}, Pollux does better with P_2 than P_1. A strategy which dominates every other strategy is, of course, obviously the best strategy, since it is *unconditionally* at least as good (or possibly better) than any other by definition of domination. Neither of the two strategies which we have singled out as best for each of our two players, respectively, dominates every other. However, strategies C_{14} and P_1 are best, respectively, against a player assumed to be "rational."

The conclusions of game theory are typically based on such an assumption. In the course of our discussion, we shall frequently subject this assumption to closer scrutiny, but for the time being we shall continue to base our arguments upon it. However, we shall note at the outset that "rationality" can have many meanings. In particular, we can speak of levels of rationality, depending on how much of the environment (including the thought processes of the other player) is taken into account in arriving at choices of action.

As an example, consider a game in which a player has a strategy available which dominates all the other strategies. Clearly a "rational player" will choose such a strategy in preference to all the others. Note that the level of rationality required for arriving at this choice is not high. One needs only to note that the strategy in question is at least as good and possibly better than all the others *regardless of what the other player does.* The principle which dictates the choice of such a strategy is sometimes called the "sure thing principle," so-called because it is unnecessary in arriving at a decision to take into account what the other player may do.

The possibility of the existence of dominating strategies in games suggests a classification of games according to the level of rationality involved in their "solution," i.e., the discovery of strategies to be prescribed as rational, namely games in which (1) each player has a dominating strategy, (2) only one of the players has a dominating strategy, and (3) neither of the players has a dominating strategy.

An example of a game in which both players have dominating strategies is shown in Game 3.

Clearly Castor's dominating strategy is C_2, which gives him a larger payoff than either of the others regardless of Pollux' choice. Similarly Pollux' dominating strategy is P_1. (Recall that Pollux' payoffs are the negatives of Castor's.)

	P_1	P_2	P_3
C_1	−3	0	1
C_2	−1	5	2
C_3	−2	2	0

Game 3

An example of a game in which only one of the players has a dominating strategy is shown in Game 4.

	P_1	P_2	P_3
C_1	−3	0	−10
C_2	−1	5	2
C_3	−2	−4	0

Game 4

Here Castor still has a dominating strategy, C_2. But none of Pollux' three strategies dominates either of the others. We see that P_1 is better than P_2 against Castor's C_1 and C_2 but not against C_3; P_1 is better than P_3 against C_2 and C_3 but not against C_1; P_2 is better than P_3 against C_3 but not against either of the two others; etc.

In this game, therefore, although Castor need not consider what Pollux will do, Pollux must consider what Castor will do. Since Castor has a dominating strategy, what he will do (if he is rational) is obvious; he will choose the dominating strategy. On the basis of *that* assumption, it is quite clear what Pollux should do, namely choose P_1, which is best against C_2.

Finally we may have a game in which neither player has a dominating strategy. An example of such a game is shown in Game 5.

	P_1	P_2	P_3
C_1	−3	18	−20
C_2	−1	5	2
C_3	−2	−4	15

Game 5

Here it is no longer obvious to either player what strategy should be chosen. Nor can either player decide on obvious grounds what the other is going to choose.

Let us try to arrive at a rational decision by reasoning about the strategic structure of this game. Let us select some provisional principle of choosing a strategy. A number of such principles come to mind. One might, for example, choose the strategy which offers the possibility of the largest payoff. For Castor this would be the row which has the biggest entry in it. Another possible principle would be to choose the strategy with the largest average payoff (assuming that the choices of the other player are equiprobable).

It should be apparent that neither of these principles satisfies the criterion of maximizing one's own payoff under the constraints of the game. For the single crucially important constraint on the game is the rationality of the opponent. One should therefore assume that whatever one has figured out to be to one's advantage, the opponent has also figured out; and he may be able to prevent the hoped-for outcome. Respect for the opponent's perspicacity is a typical example of rationality in games of strategy. Experienced chess players, playing against able opponents, usually do not build their strategies around plans in which a crucial element is the expectation that the opponent will not see the traps set for him. In short, a rational player does not expect stupid decisions from his opponent.

Suppose now Castor, playing Game 5, chose by the

first principle, namely C_1, where his biggest payoff is entered (C_1, P_2). He must now ask himself "What would Pollux do if he knew what I am about to choose?" The answer is obvious. On the basis of Castor's row *1*, Pollux would P_3. The resulting payoff is a loss of *20* to Castor. Surely Castor can do better than that.

Let us now suppose that Castor chooses the strategy with the largest average payoff, namely C_3, the average being computed by assuming that Pollux' choices are equiprobable. This choice is already questionable on the grounds that there is no reason to suppose that Pollux will choose at random from his three strategies. Nor does it help to assign probabilities to Pollux' possible choices without somehow justifying these probabilities on grounds other than one's belief or hunch. Let us, however, pursue the implication of the policy just assumed. Consider what would happen if Pollux knew that Castor is choosing the row with the greatest average payoff to him. Knowing this, Pollux would choose P_2 to win 4 from Castor. As we shall see, Castor can do better than this.

From the game matrix (Game 5) we see that Castor has a *guaranteed* payoff, namely -1, which he is sure to get (possibly more) if he chooses C_2. Observe that there is a "worst" outcome for Castor in each of the rows. (We assume that if several outcomes are "equally worst," any one of them can be taken to be the worst.) Suppose Castor were to choose the row which contains the best of these worst outcomes. (Again we assume that if several outcomes are "equally best," any one of them can be called the best.) Then no matter what Pollux did, Castor would be sure to get at least this best of the worst. Next, suppose Pollux assumes that Castor will do just that. Specifically, suppose that Pollux has assumed that Castor has chosen C_2. Clearly Pollux should on the basis of this assumption choose P_1. Now knowing that Pollux, acting on the assumption that Castor has chosen C_2, has chosen P_1, Castor must choose C_2, which is best against P_1.

In this way, the joint choice (C_2, P_1) appears as a sort of equilibrium or "balance of power." The payoff in that entry is the best that either of the players can do, given that he is playing against a rational opponent.

An entry such as (C_2, P_1) is called a *saddle point* of the game matrix. The term is suggestive. The center of a saddle is the lowest point on the horse's back in the horse's longitudinal plane (i.e., as one moves from front to back) and at the same time the highest point in the plane perpendicular to the horse's motion (i.e., as one slides from side to side). Hence the saddle point is at the same time a minimum and a maximum. The term "minimax" (or "maximin"), frequently used in game theory, is in this context synonymous with saddle point.

A game matrix may have more than one saddle point. An example is shown below:

	P_1	P_2	P_3	P_4	P_5
C_1	10	2	3	2	3
C_2	−5	0	1	−7	−4
C_3	−4	−1	8	−7	10
C_4	4	2	7	2	5
C_5	6	−3	5	0	0

Game 6

Here the entries (C_1, P_2), (C_4, P_2), (C_1, P_4), and (C_4, P_4) are saddle points. ("Saddle point" will refer in our discussion both to the row-column position and to the associated payoff.) Take (C_1, P_2) and (C_4, P_4), which are both in different rows and in different columns. Clearly they cannot be the *only* saddle points in a game matrix, if the entries in them are equal. For if these were the only saddle points, then the entry in (C_1, P_4) would have to be at least as large as the saddle point entry in (C_1, P_2) and at most as large as the saddle point entry in (C_4, P_4). But the entries in (C_1, P_2) and (C_4, P_4) are equal, and therefore must be equal to the entries in (C_1, P_4) and in (C_4, P_2). Suppose now, two saddle points in different rows and different columns were not equal. Let a_{ij}, the entry in the i-th row and j-th column, be smaller than a_{kh}, the entry in the k-th row and h-th column. Then the entry a_{kj} must be at least as great as a_{kh}, being in the same row of which a_{kj} is minimal. But then a_{kj} is larger than a_{ij}, contrary to the assumption that a_{ij} is a saddle point and hence maximal in the j-th column.

We are thus led to the conclusion that if (i, j) and (k, h) are both saddle points, then their entries must be equal, and moreover (i, h) and (k, j) are also saddle points with entries equal to those in the other two. In short, when we speak of "the" saddle point of a game matrix, we may mean one of the (equal) entries of any of its saddle points.

The minimax principle can now be restated as follows: *If a two-person zero-sum game has saddle points, the best each player can do (assuming both to be rational) is to choose the strategy* (i.e., the row or column of the game matrix) *which contains a saddle point.*

We have seen that if the game matrix has several strategies containing saddle points, then regardless of how the players choose their strategies, the outcome (in the sense of payoffs) will be the same: namely the payoffs will be entered in the saddle points, which are all equal.

Let us return to the game matrix (Game 2) of Square-the-Diagonal. We see six saddle points, namely, (C_{12}, P_1), (C_{13}, P_1), (C_{14}, P_1), (C_{12}, P_4), (C_{13}, P_4), and (C_{14}, P_4). These are saddle points, because their entries are both the smallest in the corresponding rows and the largest (taking sign into account) in the corresponding columns. (Recall the smallest and largest means any one of the smallest or largest.) Note that there are several other boxes of the game matrix with entry 9, but none of the others is a saddle point.

According to the minimax principle, then, Castor should choose either C_{12}, or C_{13}, or C_{14} (it doesn't matter which), while Pollux should choose P_1 or P_4 (it does not matter which).

Let us nevertheless compare C_{12}, C_{13}, and C_{14}. We find that C_{14} dominates the other two. Therefore from a certain point of view C_{14} should be preferred to C_{12} and to C_{13}. We see that the choice of C_{14} by Castor allows him to cash in on a possible mistake by Pollux, for example, if Pollux should (foolishly) choose P_3, P_5, or P_7. However, if we continue to assume the players' rationality this additional consideration should play no part in choosing a strategy among those which contain a saddle point.

Comparing P_1 and P_4, we see that neither dominates the other. For example, P_1 is better than P_4 against C_4, but P_4 is better than P_1 against C_5. Therefore there is no additional consideration (such as providing for Castor's possible mistakes) that can be conclusive for Pollux' choice between P_1 and P_4. If Castor is not expected to make a mistake, the choice between P_1 and P_4 is indifferent. In order to make his choice, if he has reason to suppose that Castor will make a mistake, Pollux must know what sort of mistake Castor will make. We must keep in mind that the choice of *strategy* (unlike the choice of sequential moves) is made *before the game starts* and that by definition of strategy no information obtained in the course of play in a game is relevant to changing the strategy in the course of the game.

The minimax principle constitutes a general solution of all two-person zero-sum games with saddle points. It is not a practical solution; it does not determine specific prescriptions of how to play specific games. To obtain such a practical solution for a given game one must first reduce the game to normal form. In view of the enormous number of strategies available even in most of the simplest games, this is usually a physically impossible task. The value of the general solution is conceptual rather than "practical." It tells us that if a game has a saddle point, there *is* a best strategy from the standpoint of each player. If each player finds this best strategy, the outcome of every play of the game is determined in advance. In a sense, therefore, there is no point in playing such a game. Tic-Tac-Toe is a game of this sort.

It is also shown in game theory that all the so-called "games of perfect information" have saddle points. In these games at each move each player knows exactly the position reached in the game so far. Chess is a game of perfect information. Therefore if the best strategy available to White and to Black respectively were ever found, the outcome of every game of Chess would be the same and known in advance. In fact, even today most Chess games are broken off before they are formally ended, when it becomes clear that one of the players (or neither) has a winning strategy at his disposal.

Poker, on the other hand, is not a game of perfect information, because no player knows with certainty what cards are held by the others. Formally speaking, no player knows the outcome of a move made by a fictitious player called Chance, who at the beginning of each deal chooses among all the possible arrangements of the deck.

Games which are not games of perfect information may not have any saddle points, and so the minimax principle in the form we have stated it need not apply to such games. In the next chapter we shall consider an extension of the minimax principle which does apply to zero-sum games without saddle points.

6. *Mixed Strategy*

If, in the course of the game, some player's choice on a given move remains unknown to the other players *after the move is made,* then the game is no longer a game of perfect information. A variant of Chess called Kriegspiel is a well known example. As mentioned earlier, the rules of this game are the same as those of Chess except that each player knows only the positions of his own pieces, because his opponent's moves are kept secret. Kriegspiel requires the participation of a referee, whose function is to remove captured pieces, to declare check, and to forestall moves which, unbeknown to the player who makes them, are illegal (such as moving into check).

Let us return to the Square-the-Diagonal game. Suppose that each player makes his choice *in ignorance* of the other's preceding choice. Under these conditions, Castor can no longer choose strategies in which the last choice is made conditional on Pollux' choice. The only strategies available to Castor are those in which he decides in advance both of his choices (recall that Castor moves twice) without regard for Pollux' choice, since without knowledge of Pollux' choice it is impossible to take that choice into account. It follows that the only strategies available to Castor are C_1, C_4, C_7, C_{10}, C_{13}, C_{16}, C_{19}, C_{22}, and C_{25}. Note that those are the strategies in which Castor's second choice does not depend on Pollux', a fact expressed by the identity of the last two numbers in the triples designating Castor's strategies. For example, C_4 says "Start with 1; following Pollux' choice, choose 2 *regardless* of how Pollux has chosen." In this context, "regardless" is synonymous with "in ignorance of."

Pollux' situation is similar. Where he formerly had eight available strategies, he now has only two, namely P_1 and P_8, since for him to choose in ignorance of Castor's choice means to choose 1 or 2 unconditionally, and this is just what strategies P_1 and P_8 prescribe.

The game matrix of Square-the-Diagonal has now shrunk. It has been reduced from twenty-seven rows and eight columns to nine rows and two columns (see Game 7).

It turns out, however, that this Kriegspiel version of Square-the-Diagonal still has a saddle point, namely at (C_{13}, P_1). We shall soon see that we cannot expect this to happen in general. Typically, a game loses its saddle point when it is no longer a game of perfect information. Our example is instructive, however, in showing that even a game which is not a game of perfect information may have a saddle point. We thus learn the following fact about Square-the-Diagonal: neither player has anything to gain from knowing how the other has moved, if both are rational. Therefore the game may just as well be played with the "cards on the table." If the game were a paradigm of some military situation, military secrecy would be pointless in this case.

Let us now turn our attention to another game, in which the presence or absence of information makes a

	P_1	P_8
C_1	−3	−6
C_4	−6	9
C_7	−11	−14
C_{10}	−6	9
C_{13}	9	12
C_{16}	−14	17
C_{19}	−11	−14
C_{22}	−14	17
C_{25}	−19	−22

Game 7

crucial difference. This is a variant of Button-Button, a game one plays ordinarily only with babies but which is quite instructive in some respects.

Button-Button is played as follows. Hider conceals a button in either hand, and Guesser tries to guess in which hand the button is concealed.

Trivial as this game seems, we shall make it even more trivial by making Hider's choice known to Guesser. To compensate for this simplification, we shall add a complication, namely different payoffs to Hider depending on all the four possible outcomes, instead of only on two (guessed and not guessed.) The four outcomes are (1) button in left hand, and Guesser says "Left"; (2) button in left hand and Guesser says "Right"; (3) button in right hand and Guesser says "Left"; (4) button in right hand and Guesser says "Right."

The game tree is shown in Figure 3, the payoffs being

those to Hider. As before, the game is zero-sum, so that the payoffs to Guesser are the same with the opposite sign.

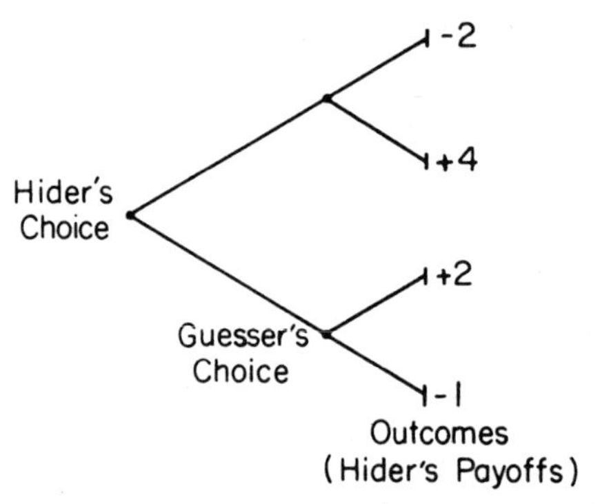

FIG. 3. Button-Button in extensive form.

We can now reduce the game to normal form. Hider has only two strategies, namely H_1 (left) and H_2 (right). Guesser, however, has four strategies, namely G_1 (guess left regardless of where the button is; G_2 (guess the hand where the button is); G_3 (guess the hand where the button is not); G_4 (guess right regardless where the button is). Note that to make the analysis complete we must list *all* the strategies, even the obviously foolish ones like G_3.

The matrix is shown as Game 8.

	G_1	G_2	G_3	G_4
H_1	−2	−2	4	4
H_2	2	−1	2	−1

Game 8

We observe a single saddle point, namely, at (H_2, G_2). The prescription, to Hider, therefore is to hide the button always in the right hand, and to Guesser to guess the hand in which the button is. This is, of course, the prescription dictated by common sense. It is obviously best for Guesser, and it is also best for Hider (under the circumstance that Guesser always knows where the button is) who stands to lose less by putting the button in his right hand. However, let not the utter triviality of this game and its solution obscure its theoretical importance. For if we now deny to Guesser information as to where the button is, the game ceases to be trivial.

Observe that in the absence of this information only two strategies are available to Guesser, namely G_1 and G_4, for he can now guess only unconditionally left or right. The game matrix becomes as is shown for Game 9.

	G_1	G_4
H_1	-2	4
H_2	2	-1

Game 9

How shall Game 9 be played? Let us see what we can conclude if we apply in turn each of the three possible principles discussed in Chapter 5, namely choosing the strategy which contains the biggest payoff, choosing the strategy which contains the biggest average payoff (assuming equiprobability of the other's choices) and choosing the strategy which contains the best of the worst.

Suppose Hider chooses H_1, hoping for the biggest payoff (4). But if he is consistent in his choices and Guesser knows this, Guesser will choose G_1 and so the outcome will be the worst possible for Hider (-2).

Suppose Hider calculates his average payoffs. Assuming equiprobable choices by Guesser, the expected (average) payoff H_1 is $+1$ from H_1 and $+\frac{1}{2}$ from H_2. Hence H_1 is indicated. But if Guesser knows this, he can again win 2 by choosing G_1.

Suppose finally that Hider is guided by the minimax principle. His best of the worst is now in H_2, which guarantees him a payoff of -1. But if he accepts -1, he is no better off than if Guesser always knew where the button was hidden. We feel that this cannot be right. Certainly Hider must derive some advantage from denying to Guesser the knowledge of where the button is placed. How can Hider secure this advantage?

Hider's failure to secure an advantage stems from not having pursued the strategic analysis of the game to its conclusion. Let us now do so.

Suppose after Hider has *provisionally* decided to choose H_2 (for that is where his best of the worst is), he asks himself what Guesser would do if he knew of his (Hider's) decision. The answer is obvious. Guesser would choose G_4. But now knowing Guesser's choice, what should Hider do? The answer is again obvious: he should choose H_1 (not H_2). But knowing *this*, Guesser would choose G_1 (Not G_4), and knowing *this*, Hider should choose H_2 (not H_1). The process of deciding "what he would do if he knew that I know that he knows, etc." apparently has no end. And so the minimax principle seems to be refuted in this game or, at least, seems to lead to no conclusion.

Examining the game matrix, we see at once where the trouble lies. Game 9 has no saddle point. The minimal entry in neither row is maximal in its column. To put it in another way, Hider's minimax does not coincide with Guesser's minimax, and so the "balance of power," which characterizes the games with saddle points, does not operate.

One of the fundamental results in the theory of two-person zero-sum games is the extension of the minimax method to games without saddle points. The extension is done by introducing a new concept, namely the *mixed strategy*.

Roughly speaking, the purpose of mixed strategy is to

keep the opponent guessing about what one will do. It may appear that keeping the opponent guessing (i.e., denying information to him) is always advantageous. However, as we have already seen, this is not always the case. If a game *has* a saddle point and if the opponent is rational, there is no advantage in denying information to him. For in choosing a minimax strategy one does what the opponent expects one to do, on the assumption that the opponent is rational, which in the context of game theory always includes the assumption that the opponent assumes *his* opponent (namely oneself) to be also rational. On the other hand, doing the unexpected, i.e., departing from the minimax strategy (in a game with a saddle point) can never improve and, in general, will impair the payoff of the player who does this (assuming the opponent to stick to the minimax strategy). *Theoretically*, therefore, one should never depart from a minimax strategy in a game of perfect information, like, say, Chess.

There are, to be sure, famous examples in the world of Chess of "brilliant and unconventional play." On the face of it, these examples can be construed as departures from prudent strategies and so, perhaps, from minimax strategies, since in the light of what has been said, minimax strategies are certainly prudent. In response, two things can be pointed out. First, the minimax strategies of Chess (comprising the entire game, not just a part of it, such as the end game) are unknown. Therefore, "unconventional" play may not actually be a departure from minimax strategy. Second, the minimax strategy principle is prescribed against a rational opponent, namely one who himself uses a minimax strategy. Many a "brilliant" play, even of the greatest masters, has been shown upon analysis to have been unsound, i.e., would have led to a lost game if the opponent had played differently. This is prima facie evidence that such play may have departed from minimax strategy. But the fact that it succeeded is evidence that the opponent *also* departed from minimax strategy and therefore was not rational. Minimax strategy is not prescribed against an irrational opponent.

If, then, the rationality of both players is assumed, definite strategies are prescribed for both in games with saddle points, and there is no way of confusing the opponent to one's own advantage.

The situation is entirely different in games without perfect information, such as Button-Button. In the simplest cases, where the outcomes are either "guessed" or "not guessed," and the payoffs are symmetric, it is clear that no consistent pattern of hiding the button is of advantage to Hider. For if this pattern is discerned by Guesser, Guesser can always guess. Hider must *confuse* Guesser. He can do this by *randomizing* his choices of right and left. Moreover, he must randomize them in such a way that the choices are equally probable, for if one hand is favored over the other and Guesser finds out which, Guesser can win more than he loses by *always* guessing the favored hand.

It may appear that by cleverly shifting the favored hand, Hider can entice Guesser to guess wrong most of the time. But if such tactics are successful, this merely shows that Hider is more clever in anticipating Guesser's choices than Guesser is in anticipating Hider's choices. Game theory cannot assume that one player is more clever than another. The reasonable conjecture, therefore, is that Hider should choose each hand with equal probability, but in completely random fashion; and similarly, Guesser should choose each hand randomly with equal probability. This way of playing illustrates the principle of mixed strategy, in particular the $(\frac{1}{2}, \frac{1}{2})$ strategy. (The notation $[\frac{1}{2}, \frac{1}{2}]$ indicates the respective probabilities of the two choices.) The $(\frac{1}{2}, \frac{1}{2})$ strategy can be realized by tossing a coin before each choice and by making the choice dependent on the outcome of the toss. Here it is of the essence to conceal one's *specific* choices from the other. But in another sense, no secrecy is involved, because each player can assume that the other is choosing each hand half the time. If this assumption is correct, the *mixed* strategy is not a secret. The assumption can be made wrong only by departing from the $(\frac{1}{2}, \frac{1}{2})$ strategy, which is a disadvantage to the player making the departure.

When the winning and losing payoffs are not numerically equal, as in Game 9, it is no longer true that the $(\frac{1}{2}, \frac{1}{2})$ mixed strategy is best for either player. To show this, let us calculate the long-range (expected) gains or losses accruing to each player if each uses the $(\frac{1}{2}, \frac{1}{2})$ mixed strategy.

Since the players make their choices independently, each of the four outcomes will occur with probability $\frac{1}{4}$. This means that Hider can expect on the average a payoff of $(-2)\frac{1}{4} + (4)\frac{1}{4} + (2)\frac{1}{4} + (-1)\frac{1}{4} = \frac{3}{4}$ units per play and, of course, the guesser must expect to lose this amount. However, as we shall see in a moment, Guesser can reduce the expected gain of Hider, and there is nothing Hider can do about it.

Let Guesser favor G_1 slightly, namely guess the left hand five times out of nine. Then every time Hider chooses the same hand (H_1), Guesser will win 2. This will be 5/9 of the time. Consequently, Guesser's expected gain in this outcome is $2 \times 5/9 = 10/9$. Guesser will also guess wrong 4/9 of the time (when Hider chooses H_1), and his loss will be $4 \times 4/9 = 16/9$. Consequently Guesser's expected gain when Hider chooses H_1 will be $10/9 - 16/9 = -2/3$.

If, on the other hand, Hider chooses H_2, Guesser will be wrong 5/9 of the time, which will give him $-2 \times 5/9 = -10/9$; and he will be right 4/9 of the time, which will give him $1 \times 4/9 = 4/9$. Consequently when Hider chooses H_2, Guesser will have an expected gain of $-10/9 + 4/9 = -2/3$, exactly as before.

It follows that *no matter what Hider does,* he cannot *expect* to win more than $\frac{2}{3}$ per play (in the long run) if Guesser follows the policy just described. Guesser can therefore do better with a (5/9, 4/9) strategy, which gives him an average loss of $\frac{2}{3}$ per play, than with a ($\frac{1}{2}$, $\frac{1}{2}$) strategy, which gives him an average loss of $\frac{3}{4}$ per play.

We have seen that Guesser can get an expected pay-off per play of $-\frac{2}{3}$ no matter what Hider does. Hider, however, must also protect himself with a mixed strategy, for if he chooses H_1 or H_2 consistently, Guesser can take advantage of this and win 2 in the one case or 1 in the other (instead of losing $\frac{2}{3}$ on the average).

Let Hider, therefore, choose the mixed strategy ($\frac{1}{3}$, $\frac{2}{3}$), i.e., let him favor H_2 (right hand) over H_1 two-to-one. Then, when Guesser chooses G_1, Hider can expect to win $(-2)1/3 + (2)2/3 = -2/3 + 4/3 = 2/3$. When Guesser chooses G_2, Hider will win $(4)1/3 + (-1)2/3 = 4/3 - 2/3 = 2/3$, exactly the amount expected to be lost by Guesser. Hider cannot improve his expectation any more than Guesser can.

Now take some other mixture. Suppose, for example, that Hider mixes H_1 and H_2 strategies in proportion ($\frac{1}{4}$, $\frac{3}{4}$). We now ask whether Guesser can find a strategy mixture which will give Hider less than an average of $\frac{2}{3}$ per play. Let $(y, 1 - y)$ be such a mixture. Then Hider's expected win will be

$$y[(-2)(1/4) + (2)(3/4)] + (1 - y)[(4)(1/4) + (-1)(3/4)] = (1 + 3y)/4 \qquad (2)$$

We see that if $y < 5/9$ Hider's expected gain necessarily will be less than $\frac{2}{3}$.

This happened when the Hider chose Right with greater frequency than $\frac{2}{3}$. Let us now see what may happen if he chooses Right with a frequency smaller than $\frac{2}{3}$, say three times out of five. Let us see whether now Guesser has a mixture which will give Hider less than $\frac{2}{3}$ expected win. Again call this mixture y. The Hider can now expect

$$y[(-2)(2/5) + 2(3/5)] + (1 - y)[(4)(2/5) + (-1)(3/5)] = (5 - 3y)/5 \qquad (3)$$

Now we see that Hider's expected gain can be made less than 2/3 if $y > 5/9$.

Suppose now Hider mixes his strategies in any other proportion than $(\frac{1}{3}, \frac{2}{3})$. Call this proportion $(x, 1 - x)$. Then, assuming that Guesser mixes his strategies in proportion $(y, 1 - y)$, we have the following formula for Hider's expected gain:

$$y[(-2)x + 2(1 - x)] + (1 - y)[(4)x + (-1)(1 - x)] = y(3 - 9x) + 5x - 1. \qquad (4)$$

Whenever $x < \frac{1}{3}$, this expression can be made smaller than $\frac{2}{3}$ by choosing y sufficiently small. Whenever $x > \frac{1}{3}$, Hider's payoff can be made smaller than $\frac{2}{3}$ by choosing y sufficiently large. Only when $x = \frac{1}{3}$, the choice of y cannot affect the resulting payoff $(\frac{2}{3})$, because in that case the coefficient of y, namely $(3 - 9x)$ vanishes.

On the basis of these results, the game theoretician concludes that the mixture $(\frac{1}{3}, \frac{2}{3})$ is the best mixture for Hider. It gives him a *guaranteed* expected gain of $\frac{2}{3}$, and no other mixture does this. Similarly, the best mixture for Guesser is (5/9, 4/9). This completes the analysis of the game. Rational mixed strategies can be prescribed to both players as follows: Hider should hide the button in one or the other hand without any pattern, i.e., completely randomly, but choosing the right hand twice as frequently on the average as the left. Guesser should likewise randomize his choices, guessing the left hand 5/9 of the time. As a result, Hider will win on the average (in the long run) $\frac{2}{3}$ per play, and, of course, Guesser must lose this amount. Nothing that Hider can do can increase his expected gain of $\frac{2}{3}$. The outcome of each play is not determined, but the long-run average outcome is.

This result is generalizable. A similar statement can be made about any two-person zero-sum game. In other words, once the strategies available to each player have been listed and the payoffs entered, it is possible in principle to indicate to each player how to mix the available strategies so as to obtain an expected gain (positive or negative) which cannot be improved upon as long as both players are rational. "Rational" in this context means seeking to maximize one's own expected gain under the constraints of the game. The constraints, be it noted, are the other player's attempts to do the same thing.

In the strategy mixture of a game with a saddle point probability one (certainty) is assigned to a strategy containing the saddle point and probability zero to every other. In other words, the strategy mixture is replaced by a pure strategy. Thus every two-person zero-sum game has a pair of strategies either pure or mixed which are optimal in the sense described above.

The Idea of General Solution

The result we have derived constitutes the general solution of every two-person zero-sum game. A discussion of the sense in which the game theoretician speaks of a solution may serve to clarify the nature of game theory and, perhaps, to remove some misconceptions about it. Let us examine a somewhat analogous situation in mathematics.

There is an ancient Egyptian treatise on mathematics known as the Rhind Papyrus (dating from ca. 1700 B.C.). The treatise contains a collection of problems which involve the solution of algebraic equations. In particular, it contains this brain teaser:

> Divide 100 loaves among 5 men so that the shares are in arithmetic progression and so that the sum of the larger three shares is seven times the sum of the smaller three shares.

To the modern mathematician such problems are utterly uninteresting. He recognizes the *type*, namely a problem formulated as two linear equations in two unknowns. The modern mathematician can write down the solution of every such pair of equations which will include the solutions of *all* such problems regardless of the number of men or loaves or of the ratio of the sums.

In the formulation of the general problem, the number of men, the number of loaves, the number of smaller shares (and therefore of the larger shares) and the ratio of the sums of the shares are assumed known. These are the *parameters* of the problem. The size of the smallest share and the difference between successive shares are unknown. These are the variables of the problem. The pair of equations is

$$\begin{aligned} Ax + By &= C \\ Dx + Ey &= F \end{aligned} \tag{5}$$

where x is the smallest share, y is the difference between shares, while A, B, C, D, E, and F are expressions involving the parameters. The general solution of the problem is

$$x = \frac{CE - BF}{AE - DB} \tag{6}$$

$$y = \frac{AF - DC}{AE - DB} \tag{7}$$

If the parameters are known, A, B, C, D, E, and F can be calculated. When these are known, x and y can be calculated.

For example, in the context of the Egyptian problem, if x is the number of loaves in the smallest share and y the difference between successive shares, then it can be shown that $A = 5$, $B = 10$, $C = 100$, $D = 11$, $E = -2$, $F = 0$. Solving the resulting equations, we obtain $x = 1\frac{2}{3}$, $y = 9\frac{1}{6}$. The shares are thus $1\frac{2}{3}$, $10\frac{5}{6}$, 20, $29\frac{1}{6}$, $38\frac{1}{3}$.

The "practical" man is not impressed by the general

form of the solution given by (6) and (7). He wants an answer to a specific problem, the actual numbers of loaves received by each man. In principle he can get the numbers from the general solution, but he cannot use the solution "in principle."

The mathematician, on the other hand, is interested in what can be said *about* the solution, not in the specific solution. For example, from (6) and (7) he can conclude that if $AE = DB$, while $CE \neq BF$ and $AF \neq DC$, the problem has no solution at all. He can conclude that if $AE = DB$, $CE = BF$, $AF = DC$, the problem has an infinity of solutions; also that under certain conditions the smallest share will be negative which may or may not be an acceptable solution, depending on whether the "loaves of bread" are supposed to be physical loaves or merely bookkeeping entries, and so forth.

The example is meant to illuminate the difference in outlook between someone faced with a specific problem and a mathematician who is usually interested in the most general formulation of *classes* of problems. Sometimes the mathematician comes up with a formula which in one stroke solves all problems of a given class, but this does not happen frequently. The mathematician's elucidation of problems sometimes leads not to a solution but to a clarification, namely of what it is that the problem involves, what obstacles stand in the way of solutions, what special cases of the problem can be treated by what methods, and the like.

Solving problems is not the only, nor even the most important task of mathematical theory. Mathematics is all of one piece—all of its parts are interconnected, sometimes by extremely subtle threads of logic. Any discoveries which shed light on one class of mathematical problems usually have relevance to other areas. The interlocking of the different mathematical developments makes the entire structure of mathematics understandable and harmonious. It is this unity of method that has made mathematics the most powerful tool of creative deductive thought known to man.

Game theory should be viewed in the same way. The focus in game theory is on the *general principles governing the logical structure of strategic conflict.* The existence of a best mixed strategy for each of the players in a two-person zero-sum game is one such principle. It is quite analogous to the existence theorems of algebra, for example, the theorem that there exist two real roots of a quadratic equation $ax^2 + bx + c$ with real coefficients, provided $b^2 \geqslant 4ac$, but not otherwise. The existence theorem says nothing about how to find the roots, but it says a great deal nevertheless; for clearly there is no sense in trying to find the real roots of such an equation if they do not exist. Similarly the expression "There's gold in them thar hills" is significant, if true, even if it gives no indication about where the gold is. Similarly Mendeleev's periodic table was significant in giving an indication of the possible existence of elements not yet discovered, although it said nothing about how they might be discovered. Similarly Einstein's equation $E = mc^2$ was significant in calling attention to the energy locked in the atom, even thought it gave no indication of how that energy might be released.

7. *Solving the Two-Person Zero-sum Game*

The existence theorem on two-person zero-sum games tells us that we shall not be attempting the impossible in searching for a best strategy for each player. It tells us also that if the game has a saddle point, then there exists among all the available strategies a best one for each player (possibly several equally good ones). If the game has no saddle point, then there is no best "pure" strategy. In that case the players should mix their strategies. A best mixture (possibly several) is available among the possible strategy mixtures. However, what the existence theorem does not tell us is how to find the best strategy mixture. Let us see how we might proceed in the case of the 2×2 game.

Game 10 is a general game of this sort, in which the payoffs (always assumed to be the row chooser's) may be anything.

Suppose we classify all such games depending on the ways the four payoffs to Castor are arranged in the order of magnitude, for example, $a > b > c > d$, $a > b > d > c$, $a > c > b > d$, etc. Next suppose a is a saddle point.

	P_1	P_2
C_1	a	b
C_2	c	d

Game 10

Then because of the way saddle point is defined, $b > a > c$. But we already have the solution of all games with saddle points. Therefore we wish to omit them from consideration. We seek solutions of games *without* saddle points, and so we must eliminate all the 2×2 games in which $b > a > c$. These are the games in which the payoffs are ordered as follows: $d > b > a > c$; $b > d > a > c$; $b > a > d > c$; $b > a > c > d$. Assuming, in turn, that b, c, or d is a saddle point, we remove all the games in which this is the case.

This procedure leaves only the games to be considered in which $a > d > b > c$ or $d > a > b > c$, or $a > d > c > b$, etc., in short where the arrangement is such that both payoffs in one of the diagonals of the game matrix are greater than either of the payoffs in the other diagonal. This means that either Castor or Pollux prefers the two outcomes in which both players use the same strategy (either 1 or 2) to the two outcomes in which they use disparate strategies. A game of this sort is seen to be similar to Button-Button, in which Guesser always prefers the pair of outcomes in which both players have chosen the same hand, and, of course, Hider prefers the other pair of outcomes. Clearly it does not matter how we label the players or the outcomes. Therefore we can represent all the 2×2 games without saddle points by Game 10, in which it is understood that $a \geqslant d > b \geqslant c$. Note: we must assume that $d > b$ (i.e., that the inequality is strict), otherwise b would be a saddle point.

The existence theorem tells us that each player can guarantee himself a certain minimum expected payoff if he uses a certain mixed strategy. Let v be the payoff to Castor if both players use the prescribed mixed strategies. The quantity v is called the *value* of the game. We shall have *solved* the game if we find all the mixed strategies $(x, 1-x)$ prescribed to Castor and all the mixed strategies $(y, 1-y)$ prescribed to Pollux, such that when both use such mixed strategies, Castor can expect the payoff v, and consequently Pollux can expect $-v$.

We proceed to find x and y.

Since v is guaranteed to Castor, Castor should get at least v regardless of how Pollux plays. If Pollux plays P_1, Castor can expect $ax + c(1-x)$, and this amount must not be less than v. Therefore we write

$$ax + c(1 - x) \geqslant v. \tag{8}$$

Analogously we can write (assuming Pollux plays P_2)

$$bx + d(1 - x) \geqslant v. \tag{9}$$

Since Pollux should get at least $-v$, regardless of how Castor plays, we can also write[8]

$$ay + b(1 - y) \leqslant v; \tag{10}$$

$$cy + d(1 - y) \leqslant v. \tag{11}$$

Finally, since x and y are probabilities, we must have

$$0 < x < 1; \qquad 0 < y < 1. \tag{12}$$

The last inequalities are strict, since otherwise we would allow the existence of optimal pure strategies, which are ruled out by the absence of a saddle point.

We shall have found x, y, and v if we find three corresponding numbers which satisfy all of the conditions [(8)-(12)].

Note that if we find values which satisfy some of the conditions with the inequality signs replaced by equality signs, these values will do, because the condition "greater than *or* equal to" includes "equal to." Let us therefore try to solve some *equations* which result if the inequality sign is replaced by an equality sign. We choose (8) and (9) and accordingly write

$$ax + c(1 - x) = v; \tag{13}$$

$$bx + d(1 - x) = v. \tag{14}$$

Since the right sides of these equations are equal, we can write

$$ax + c(1 - x) = bx + d(1 - x). \tag{15}$$

Solving for x, we obtain

$$x = \frac{d - c}{(a + d) - (b + c)}. \tag{16}$$

Solving for y, we have

$$y = \frac{d - b}{(a + d) - (b + c)}. \tag{17}$$

Since there is no restriction on v, the value of v determined by x and y reveals itself automatically as the value of the game (to Castor). It remains, however, to test the condition $0 < x < 1$ and $0 < y < 1$ and also whether the value of v determined by (13) and (14) satisfies (10) and (11).

Note that because of our arrangement $a > d > b > c$, both the numerators and the denominators on the right sides of (16) and (17) are positive. To satisfy (12), one must show that the numerators are smaller than the denominators, namely

$$d - c < a + d - b - c, \tag{18}$$

$$d - b < a + d - b - c, \tag{19}$$

in other words

$$0 < a - b, \tag{20}$$

$$0 < a - c. \tag{21}$$

But inequalities (20) and (21) must be satisfied, since $a \geqslant d > b \geqslant c$ implies $a > b$ and $a > c$. Therefore (12) is satisfied. Game 10 is now solved.

Applying exactly the same procedure to our Button-Button game (Game 9), we arrive at the following solution:

$$x = \frac{-1 - 2}{(-3) - 6} = 1/3, \tag{22}$$

$$y = \frac{-1 - 4}{(-3) - 6} = 5/9, \tag{23}$$

which was the solution found earlier.

Two things are noteworthy about 2×2 games without saddle points. First, the mixed strategy solution is unique; that is, only one mixed strategy is prescribed to each player. Second, if one of the players plays the prescribed mixed strategy (the minimax), he gets the same expected payoff whatever the other player does. These features are consequences of the fact that if a 2×2 game has a saddle point, then necessarily at least one of the players must have a dominating strategy. From this it follows that if neither player has a dominating strategy, a 2×2 game cannot have a saddle point. In larger games neither condition necessarily obtains. Some strategies may dominate others without dominating *all* the others. Also a saddle point may exist without either player having a dominating strategy. Therefore, in larger games (with more than two strategies available to each player) the conditions which insure the nonexistence of dominating strategies or saddle points can become exceedingly involved; at least it takes involved expressions to write them down. Accordingly the task of determining minimax mixed strategies in a general two-person game is usually quite tedious. The explicit methods of solving such games involve search procedures; the object of the search is essentially to eliminate from the game those strategies or strategy mixtures which are dominated by other strategies or strategy mixtures. When all such sets of strategies have been eliminated, the remaining reduced game gives rise to solutions in terms of strategy mixtures from which the eliminated strategies are missing.

To illustrate the method, let us solve a game just one step removed from a 2×2 game, namely Game 11, which is a 2×3 game.

	P_1	P_2	P_3
C_1	a	b	c
C_2	d	e	f

Game 11

We seek a mixed strategy $(x, 1-x)$ for Castor and a mixed strategy $(y, z, 1-y-z)$, for Pollux.

The inequalities to be satisfied are the following:

$$ax + d(1-x) \geqslant v, \tag{24}$$

$$bx + e(1-x) \geqslant v, \tag{25}$$

$$cx + f(1-x) \geqslant v, \tag{26}$$

$$ay + bz + c(1-y-z) \geqslant v, \tag{27}$$

$$dy + ez + f(1-y-z) \geqslant v, \tag{28}$$

$$0 \leqslant x \leqslant 1, \tag{29}$$

$$0 \leqslant y \leqslant 1, \tag{30}$$

$$0 \leqslant z \leqslant 1, \tag{31}$$

$$0 \leqslant y + z \leqslant 1. \tag{32}$$

Note that the last four inequalities permit some of the strategies to be eliminated.

We could attempt to solve a game of this sort in general, that is, in terms of the payoffs designated by letters. However, this would involve a tedious enumeration of conditions which must be satisfied if the solutions so obtained could be accepted. (For example, certain conditions must be satisfied if x, y, and z are to be confined to the interval between zero and one.) Instead of seeking the general solution, we shall construct a numerical example of a 2×3 game, keeping in mind certain features we want it to have. We shall then solve the specific game, by a method which will illustrate the general principles used in the solution of all two-person zero-sum games.

The features which we should like our game to have are those that insure (to the extent possible) "non-degeneracy." [9] In other words. we should not like our game to reduce immediately to a game with a saddle point or to a 2×2 game, since we have already found the general solution of all such games and would learn nothing new by solving a 2×3 game. In particular we do not wish either of Castor's strategies to dominate the other, since this would immediately introduce a saddle point. Next, we wish to avoid games in which one of Pollux' strategies dominates another, since in that case Pollux should simply omit the dominated strategy from his mixture. This is because increasing the "weight" (probability) of the dominating strategy at the expense of the dominated one can only improve the payoff resulting from the mixture.

One might think that a way to avoid dominating strategies in Game 11 is by making sure that no dominating strategies occur in any of its three 2×2 sub-games,[10] namely

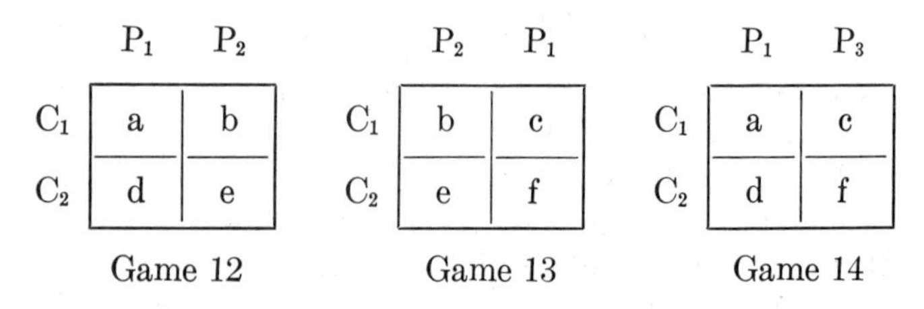

	P_1	P_2
C_1	a	b
C_2	d	e

Game 12

	P_2	P_1
C_1	b	c
C_2	e	f

Game 13

	P_1	P_3
C_1	a	c
C_2	d	f

Game 14

However, it turns out that in at least one of the sub-games Castor must have a dominating strategy, if Pollux is to have none. To see this, let us try to deny dominating strategies to either player. To insure the absence of a dominating strategy in Game 12 let us have Min (a, e) > Max (d, b); that is to say, either payoff in the first parenthesis should be larger than either in the second. We have seen above (see p. 79) that this insures the absence of a dominating strategy in a 2 × 2 game. Once we have prescribed that Min (a, e) > Max (d, b), we have no choice for Game 13 but to assign Min (e, c) > Max (b, f). We cannot assign Min (b, f) > Max (e, c), because this would contradict Min (a, e) > Max (d, b), and any other assignment would give a dominating strategy to one of the players. But the two assignments together imply Min (a, c) > Max (d, f), and so in Game 14 Castor must have a dominating strategy.

It appears, then, that the best we can do to avoid simplification (degeneracy) in our 2 × 3 game is to have none of Pollux' strategies dominate another and to have a dominating strategy for Castor in just one of the 2 × 2 sub-games (but not in the 2 × 3 game).

Game 15 is a numerical example which satisfies these conditions.

	P_1	P_2	P_3
C_1	4	0	3
C_2	−3	2	−1

Game 15

Neither of Castor's strategies dominates the other in the full game; none of Pollux' strategies dominates another.

By substituting the payoffs of Game 15 into inequalities [(24)-(28)], we obtain

$$4x - 3(1 - x) \geqslant v, \tag{33}$$

$$2(1 - x) \geqslant v, \tag{34}$$

$$3x - (1 - x) \geqslant v, \tag{35}$$

$$4y + 3(1 - y - z) \geqslant v, \tag{36}$$

$$-3y + 2z - (1 - y - x) \geqslant v. \tag{37}$$

Observe that if we can find x, y, z, and v to satisfy the five *equations* obtained from [(33)-(37)] by replacing the inequality signs by equality signs, we shall have satisfied the inequalities also, since the sign $\geqslant$ stands for greater than *or* equal to. If our values of x, y, and z furthermore lie between 0 and 1, as required by [(29)-(32)], we shall have found a solution to the game. (The question of whether there may be other solutions will remain open.) Let us therefore start by trying to solve [(33)-(37)] as a system of equations.

Before we start, we should note that the system to be solved has four unknowns, x, y, z, and v, and five equations. Such a system is called *overdetermined.* It may not be possible to solve it, because the values obtained from four of the equations may not fit the fifth. When this happens the system is said to be inconsistent.

Inspecting the system [(33)-(37)], we see that it is convenient to determine x and v from (33) and (34), then to substitute the value of v so obtained into (36) and (37) which will determine y and z. The question of consistency will then be settled by whether (35) is satisfied by the previously obtained values of x and v.

Solving (33) and (34) for x and v, we get $x = 5/9$; $v = 8/9$. Substituting 8/9 for v in (36) and (37), we obtain $y = 2/9$; $z = 7/9$. We note that x, y, and z all fall in the interval between zero and one, as required. (The fact that v also falls into this interval is merely incidental and of no significance.) Testing for consistency, we substitute 5/9 for x and 8/9 for v in (35) and get 11/9 on the left side and 8/9 on the right.

We see that the system [(33)-(37)] is inconsistent as a system of *equations.* But in asking this consistency, we have asked for too much. Only the inequalities need to be satisfied if x, y, z, and v are to be a solution of the game. The inequalities are satisfied because $11/9 > 8/9$. Therefore,

1. we have found a value of v, which satisfies the definition of the value of the game, because
2. we have also found a mixed strategy for Castor, namely (5/9, 4/9) which insures at least v to him; and one for Pollux, namely (2/9, 7/9, 0) which insures at least $-v$ to him.

We have also found that although none of Pollux' strategies are dominated by any other, still one of the strategies, namely P_3, has to be assigned weight 0 (never played). This suggests that even strategies which are not dominated should sometimes be avoided altogether, and this is revealed by game-theoretical analysis.

It is instructive to see what would have happened if we started with the system of equations derived from (34), (35), (36), and (37). Proceeding as before we would have obtained $x = \frac{1}{2}$, $y = 0$, $z = \frac{2}{3}$, $v = 1$. This would have violated inequality (33), and consequently our values for x, y, z, and v would not have served as a solution of the game. When this happens, one method of finding a solution prescribes the elimination of some of the strategies. This can be done by trial and error. Trying each strategy in turn, we would find that by eliminating P_3, we do arrive at a unique solution of the reduced 2×2 game, namely the one just found.

The general method can now be described:

1. Examine the game for saddle points. If one is found, the solution is immediate.
2. Examine the game for any strategies of either player that are dominated by other strategies. Eliminate all the dominated strategies from the game.
3. For each of the column chooser's strategies, write down an inequality which states that the row chooser expects a payoff of at least v, if he plays the mixed strategy $(x_1, x_2, \ldots x_n)$. The x's represent $n - 1$ of the un-

knowns, because only $n - 1$ of them are independent; the n-th is obtained by subtracting the sum of the others from 1.

4. For each of the row chooser's strategies, write down an inequality which states that the column player expects at least $-v$, if he plays the mixed strategy $(y_1, y_2, \ldots y_m)$. The y's represent $m - 1$ of the unknowns.

We now have a system of $(m + n - 2)$ inequalities involving $(m + n - 1)$ unknowns (v is the remaining unknown).

5. Add the conditions that all of the x's and y's must lie in the interval between zero and one, inclusive.

6. Select some $(m + n - 2)$ of the inequalities. Try to solve them as a system of simultaneous equations in as many unknowns. There are two possibilities: either the determinant of the system vanishes or it does not. If it does not vanish, the system yields a unique solution to be tested against the other criteria. If the determinant vanishes, there are again two possibilities depending on whether the system contains terms not involving the unknowns (i.e., is a nonhomogeneous system) or does not contain such terms (i.e., is a homogeneous system). In the former case the system is inconsistent. In the second case there may be an infinity of solutions.

7. If, in a solution so obtained, zero weights are assigned to some of the strategies, eliminate those strategies from the game matrix and start over.

8. If a solution without zero weights has been obtained by solving the system of equations (possibly of a reduced game), test this solution for compatibility with all the other criteria of the entire game. If it is compatible with all of them, it is a solution of the game. (There may be many solutions. For example, some weights may be any fraction within a prescribed interval.) If the system selected is inconsistent, or if it yields solutions incompatible with the other criteria, the search continues by reducing the game progressively, first by eliminating

single strategies, pairs of strategies, etc. Eventually a solution must be found, because if all but two strategies are left to each player, the reduced game must either have a saddle point or a mixed strategy solution.

9. All the strategies eliminated in the process just described are assigned weight zero in the corresponding strategy mixtures. The remaining weights are found from the solution of the first system that proves to be solvable and compatible with the remaining inequalities. The value of v is also determined in this process. We now have a strategy mixture (or a class of such mixtures) prescribed to each player, and the value of the game. These constitute the solution of the game.

We conclude this chapter by examining a game in which a *class* of mixed strategies is prescribed to each player. An example is the ancient two-person zero-sum game called Morra.

Morra has several versions of varying complexity. The simplest version is the following. Each player simultaneously shows either one or two fingers. If the number of fingers is odd one player wins, if even the other. We can see immediately that this game is strategically equivalent to Button-Button. If the payoffs are symmetrical, the strategy prescribed to each player is obvious, namely $x = \frac{1}{2}$; $y = \frac{1}{2}$. This game presents no further interest.

The next version of Morra is considerably more interesting. In this game each player shows one or two fingers and at the same time guesses the number of fingers shown by the other. If both guess correctly, or if neither does, the payoff is zero to both. If one guesses and the other does not, the payoff to the guesser is the sum of the fingers shown. We shall solve this game.

First, we construct the game matrix, as shown. Each strategy is designated by the number of fingers shown (first number of each pair) and the number guessed (second number). The payoffs are, as usual, to the row chooser.

	P_1 (1, 1)	P_2 (1, 2)	P_3 (2, 1)	P_4 (2, 2)
C_1 (1, 1)	0	2	−3	0
C_2 (1, 2)	−2	0	0	3
C_3 (2, 1)	3	0	0	−4
C_2 (2, 2)	0	−3	4	0

Game 16, two-finger Morra.

Next we write down the inequalities. It should be clear that the value of this game is zero, since the game is entirely symmetric and there is no reason why either player should be favored. This result ought to come out of the formal solution, but we may as well use the symmetry of the game to guess its value in advance. The inequalities with respect to Castor's expectations are

$$-2x_2 + 3x_3 \geqslant 0, \tag{38}$$

$$2x_1 - 3(1 - x_1 - x_2 - x_3) \geqslant 0, \tag{39}$$

$$-3x_1 + 4(1 - x_1 - x_2 - x_3) \geqslant 0, \tag{40}$$

$$3x_2 - 4x_3 \geqslant 0. \tag{41}$$

The inequalities representing Pollux' expectations are, of course, entirely analogous and so need not be written down.

We are to select three out of these four inequalities to solve as a system of three equations in three unknowns. We note at once that if both (38) and (41) are included, the complete system cannot be satisfied. For (38) and (41) jointly yield $x_2 = 0$, $x_3 = 0$. Substituting these values for the remaining two inequalities, we obtain from (39) $x_1 \geqslant 3/5$ and from (40) $x_1 \leqslant 4/7$, which is a contradiction, since $3/5 > 4/7$.

Therefore we include only one of (38) and (41) in our system. Taking (38), (39), and (40), we obtain from (38) $x_3 = \frac{2}{3}x_2$. Substituting this into (39), we obtain $x_1 = \frac{3}{5} - x_2$. Substituting both results into (40), we obtain $-x_1/3 = 0$, hence, $x_1 = 0$. But $x_1 = 0$ means that one of the strategies has been eliminated. Therefore we must start over again with a reduced game without the strategy C_1; otherwise we may not obtain all of the solutions. Note that entirely similar reasoning will eliminate the strategy P_1 from Pollux' choices. The reduced game leads to the following inequalities:

$$-2x_2 + 3x_3 \geqslant 0, \tag{42}$$

$$-3 + 3x_2 + 3x_3 \geqslant 0, \tag{43}$$

$$4 - 4x_2 - 4x_3 \geqslant 0. \tag{44}$$

$$3x_2 - 4x_3 \geqslant 0. \tag{45}$$

If equalities are substituted for inequalities in (43) and (44), we get $x_2 + x_3 = 1$. Moreover we know that $x_2 + x_3$ cannot exceed one (because $x_1 + x_2 + x_3 + x_4 = 1$). Therefore we know that both x_1 and x_4 are eliminated. This leaves only inequalities (42) and (45). We know that they cannot be satisfied as equations, because the only solution of these two equations is $x_2 = x_3 = 0$, which contradicts $x_2 + x_3 = 1$. Hence we must choose x_2 and x_3 so as to satisfy (42) and (45) as inequalities. This gives

$$3x_2/4 < x_3 < 2x_2/3, \tag{46}$$

$$x_2 + x_3 = 1. \tag{47}$$

The result is, of course, entirely analogous for the other player. In words this means that in playing the two-finger Morra game just described, one should never guess the same number of fingers that one shows. The result is not intuitively obvious, and it would be interesting to know whether experienced Morra players (the game is said to have been popular in antiquity) were aware of it. If the players follow this advice, then either both will guess correctly or neither will guess. The outcome will be zero in either case. This would make the game utterly uninteresting, but this is the only prudent way of playing it (including the prescribed limits on the frequencies of the two strategies). A departure from the prescription on the part of either player can be taken advantage of by the other.

At this point one might ask the following question. If the strategies prescribed lead only to zero payoffs, what difference does it make in what proportion they are mixed? The answer is that the strategies should be mixed in the prescribed proportions in self-defense, as it were. Any consistency on the part of one player can be taken advantage of by the other. Should, for example, Castor play C_3 exclusively, Pollux can win 4 by switching to P_4. Should Castor choose consistently C_2, Pollux can win 2 by switching to P_1.

We note that the "yield" of the prudent strategy is not high. For example, if Castor chooses the mixture $x_2 = \frac{3}{5}$, $x_3 = \frac{2}{5}$, he will gain nothing from Pollux' use of the prohibited strategy P_1 and only an average of $\frac{1}{5}$ from Pollux' use of the prohibited strategy P_4. Obviously Castor wins nothing as long as Pollux sticks to P_2 and P_3. Therefore Castor, playing the above mixture, can expect to win from an opponent, ignorant of game theory, $f/5$ points per play, where f is the fraction of plays on which the opponent plays the prohibited strategy P_4.

To win more, Castor must *himself* depart from the minimax, which, of course, involves a risk (if the opponent is also clever).

To be able to take advantage of his opponent by using the prohibited strategies, Castor must be "more perceptive" than his opponent; that is, Castor must be able to anticipate Pollux' departures sooner or more accurately than Pollux anticipates Castor's. We cannot be too emphatic in pointing out that game theory, as such, has absolutely nothing to say about how one becomes "more perceptive" than one's opponent. This question is psychological, not theoretical. Its investigation requires data and methodological tools which fall wholly outside of game theory. On the other hand it is well to recognize the role of game theory in bringing such questions to light and, above all, in separating these questions from the purely logical considerations with which game theory is concerned.

8. The Negotiated Game

Conflicting parties enter negotiations if each hopes to gain from the results. Each party will gain if, as a result of the negotiations, the parties agree on a course of action expected to lead to an outcome which is preferred by both (or all) parties concerned to the outcomes which might have obtained in the absence of the agreement. If the outcomes in question are only the various outcomes of a two-person zero-sum game, then clearly no negotiated agreement can simultaneously benefit both parties. For if one of two possible outcomes is preferred by one of the parties in such a game, the other outcome is sure to be preferred by the other in the same degree.[11]

The situation is different in a nonzero-sum game. Here there may be outcomes which are preferred to other outcomes by both players. In a negotiated settlement the players can at least agree to act in such a way that one of the possibly several outcomes preferred by both players (to other outcomes) obtain. This is, of course, possible only if the rules of the game or the nature of the situation allow the players to coordinate their actions so as to attain one of the preferred outcomes.

As a simplest example, consider the game of Button-Button played as follows. If Guesser guesses where Hider concealed the button, each of the players is paid a dollar by the House. If he does not, each player pays a dollar to the House.

If the players are not able to communicate, hardly anything can be said about what each player should do. Whether they will pay or collect seems to be a matter of chance. In this sense, the game is trivial. On the other hand, if the players are allowed to coordinate their choices, they can easily agree on where to conceal the button. Thus the coordinated game is also trivial.

The coordinated game of Button-Button is trivial, because the interests of the players completely coincide. The interesting features of nonzero-sum games derive from situations in which the interests of the players partly coincide and partly conflict. The essential purpose of negotiation is to use the common interests of the players as leverage in the settlement of their conflicting interests.

As an example, consider Button-Button with negotiation allowed, in which as before both players get positive payoffs if they agree on their choice of hand, and both get negative payoffs if they do not agree. However, we shall now assume that the positive payoffs are not the same to the two players.

To add vividness to our example, let us give the game another interpretation. The players are now Man and Woman who are negotiating the matter of choosing an evening's entertainment. The man proposes opera, the woman a prize fight. Let Man get one unit of payoff if both go to the opera, while Woman gets nothing. If both go to the prize fight, Woman gets a unit of payoff while Man gets nothing. If they go their separate ways, both get negative payoffs. Namely, if Man goes to the opera while Woman goes to the prize fight, Man gets $-a$, and Woman gets $-b$. If, on the contrary, Man goes to the prize fight, while Woman goes to the opera, Man gets −c, and Woman gets −d. This game has been nicknamed Battle of the Sexes and is discussed by R. D. Luce and H. Raiffa (1957).

At this point one might raise the reasonable objection that it makes no sense for Man to go to the prize fight alone, while Woman goes to the opera alone, since if they go separately they may as well go to their respective preferred places. However, this outcome is not entirely absurd. Suppose, for example, that after a spirited argument, in which Man insists on opera and Woman on the prize fight, they go off separately. But suppose further that enroute both change their minds, each deciding to give in to the other. Man, assuming that Woman has gone to the prize fight goes there looking for her. Consequently, Man ends up at the prize fight, which he abhors, while Woman, pursuing a similar course, ends up at the opera, with which she is bored.

The situation is depicted in Game 17.

		Woman	
		O_2	F_2
Man	O_1	1, 0	−a, −b
	F_1	−c, −d	0, 1

Game 17

It may seem that to treat the problem in the most general context, we should leave all the payoffs unspecified (i.e., denote them by letters). However, we shall take advantage of the game-theoretical assumption that payoffs are utilities. Therefore we can arbitrarily fix the zero and the unit point in the payoff scale of each player independently. This we have done in the "agreement diagonal," (O_1, O_2) and (F_1, F_2). The other payoffs have been left unspecified except that they are all negative (a, b, c, and d being all positive numbers). This is in accordance with the situation we are examining, namely where failure to agree on the choice of entertainment is worse for each player than agreement even on the less preferred outcome.

Next, we shall wish to impose the condition $\text{Min}(a, b) > \text{Max}(c, d)$. This condition also reflects a feature of our game. If agreement fails because each party insists on his own preferred choice, the punishment is more severe (for both) than it agreement fails because each insists on the choice preferred by the other. Psychologically this could be interpreted as follows. If Man ends up at the prize fight alone, where it dawns on him that Woman had repented and went looking for him at the opera, his disappointment is somewhat compensated by this evidence of her affection, and vice versa. Another way of interpreting the outcome (F_1, O_2) is in terms of an Alphonse-Gaston result. The reference is to the ancient comic strip characters who habitually get into the impasse of excessive politeness ("After you, my dear Gaston. . . . No, after you, my dear Alphonse").

In short, the outcome (O_1, F_2) represents Man's and Woman's respective *threat potentials* (what each will get if he or she does not follow the other to the entertainment of her or his choice). The outcome (F_1, O_2) represents the frustration resulting from misplaced altruism.

Finally, we shall also suppose that a, b, c, and d are all smaller than 1. This last restriction is logically unnecessary. It is made simply in the interest of simplifying some calculations below.

The question before us is what should the players do, say, in repeated plays of this game? Should they alternate between (O_1, O_2) and (F_1, F_2) so as to "split the difference"? This would seem eminently fair if the game were entirely symmetric (although we shall see below that even the notion of symmetry is not unambiguous in this context). But suppose the threat available to Man, namely b (what Woman loses if she insists on the prize fight, while Man goes off to the opera) is larger than the threat available to Woman, namely a?[12] We can imagine the following conversation.

He: "Are we agreed that (O_1, O_2) and (F_1, F_2) are both better than the other two outcomes?"

She: "We are."

He: "Therefore if I go to the opera, it is in your interest to do likewise."

She: "This is true. On the other hand, if I go to the prize fight, your best choice is to come along, too."

He: "However, I prefer the opera. Therefore I shall go there."

She: "I, on the contrary, prefer the prize fight. Therefore I shall go there."

He: "What will this get you? You will lose b units of utility. Why don't you come with me and break even?"

She: "If you persist in your stubbornness, you will lose a units, whereas you could break even if you went along with me."

He: "But you stand to lose more than I if we both persist. (Recall that we are assuming here that $b > a$.)"

She: "Look at it this way. Suppose you could get your 1 utile at the opera whether I went along or not. Then you would not even want to negotiate with me. You would simply go off to your opera. But you do *not* get a full utile if I don't come along. My good will must be worth *something* to you. You must, therefore, give me something in order to assure it."

If we feel that Woman has a point, in spite of being apparently in a weaker position (having a smaller threat), then we must decide how much that point is worth. Once we decide this, we can arbitrate the conflict, i.e., we can suggest how the difference is to be split in the interest of "fairness" (however defined).

Several approaches to problems of this sort have been proposed. Collectively they constitute so-called theories of bargaining and arbitration. These theories are clearly closely connected to game theory and can be considered to be a branch of the latter. It is important to note, however, that the "purely rational" character of game theory, as originally formulated, is modified in the theories of bargaining and especially in the theories of arbitration.[13] How and why this comes about will, I hope, become apparent from the discussion to follow.

Two-person nonzero-sum games of this sort are conveniently represented in the so-called payoff space, a two-dimensional diagram, in which the horizontal axis represents the row chooser's payoffs and the vertical the column chooser's. The game just discussed is represented in Figure 4.

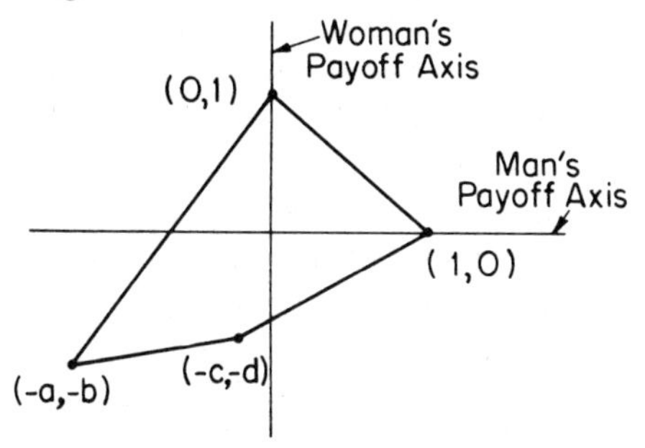

FIG. 4. The payoff polygon of the Battle-of-the-Sexes game (Game 17).

We have assumed that the two players can coordinate their choices. This means that they can mix their choices of (O) and (F) strategies in any desired proportion and moreover that each can make his choices contingent on those of the other. Thus, for example, the players can confine the outcomes to (O_1, O_2) and (F_1, F_2), mixing only these in any desired proportion. Note that if the players *cannot* coordinate their choices, they cannot attain all possible mixtures of outcomes. For example, they cannot play so as to get fifty percent (O_1, O_2) and fifty percent (F_1, F_2), because without coordination some

(O_1, F_2) and (O_2, F_1) are bound to obtain. But if they can coordinate their choices, they can certainly avoid the outcomes (O_1, F_2) and (O_2, F_1) if they so desire.

Clearly in the game we are considering the players, after reaching an agreement, will want to avoid (O_1, F_2) and (O_2, F_1). Suppose, then, they have agreed on some mixture of (O_1, O_2) and (F_1, F_2). Depending on the proportions of this mixture, the point representing the average payoff to each player will be a point on the straight line joining (1, 0) and (0, 1). The more (O_1, O_2) there will be in the mixture, the closer the point will be to (1, 0), and vice versa. In fact, it can be shown that by mixing the four outcomes in various proportions, the players can attain a pair of (average) payoffs represented by any of the points inside and on the border of the quadrangle (the payoff polygon) joining the four points shown in Figure 4.

More generally, a convex[14] polygon joining some of the points in the payoff space so as to include all of these points either on its boundary or inside, includes all of the points and only those which can be realized as payoffs by *coordinated* pairs of mixed strategies. Such a polygon is called the *convex hull* of the set of payoff points. An example of a convex hull is shown in Figure 5.

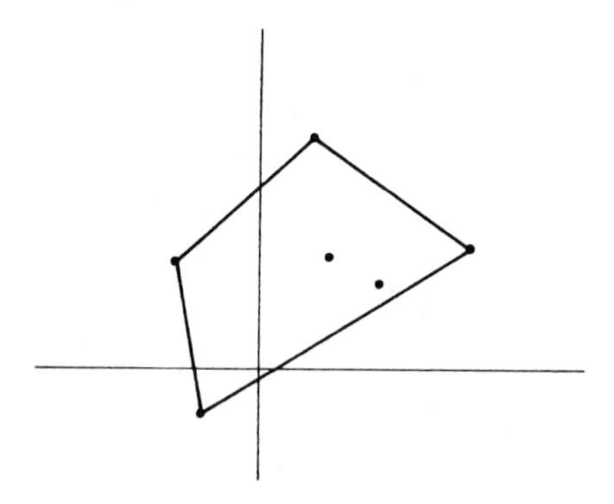

FIG. 5. A convex hull enclosing a set of payoff pairs.

We return to the Battle of the Sexes (Game 17). It stands to reason that if the players can agree on a pair of payoffs, they will not accept as a negotiated settlement any point either inside the payoff polygon or on any of its three aides except the side joining (1, 0), (0, 1). This is because *both* can get more that way than by any of the remaining points. Unless they are already on the line joining (1, 0) and (0, 1), they cannot both do better while still staying within the payoff polygon. If one does better, it will be at the expense of the other. The points on the line joining (1, 0) and (0, 1) therefore lie on the *negotiation set* of the negotiated game. Payoff pairs not in this set need not be considered by a pair of rational players.

Let us generalize the situation. Consider any two-person nonzero-sum game. All of the payoffs attainable by pure strategy pairs will be points in the payoff space. The convex hull of the space will include all of the payoff pairs which can be obtained by mixtures of *coordinated* strategies. Among these pairs will be a particular pair (x_0, y_0) which represents the respective *security levels* of the players. The security level of a player is the payoff which the player can *guarantee* to himself by choosing an appropriate mixed strategy. To obtain the security level payoff, the player need only do what he would do in a zero-sum game, in which his payoffs are the same as in the game under consideration. Clearly, a player need not accept any negotiated payoff which is less than his security level, for he can get that much without the cooperation of the other player.

Let us draw a vertical and a horizontal line through the point (x_0, y_0), i.e., the intersection of the two security levels. These lines will intersect the boundary of the payoff polygon in some pair of points. Then clearly the negotiation set must be included between these two points. Sometimes the whole portion of the boundary of the polygon so included will be the negotiation set, as shown in Figure 6a; sometimes only part of it, as shown in Figure 6b. At any rate, every point on the negotiation set must satisfy the following two conditions: (1) the players should not be able to improve their payoffs jointly from any such point, and (2) the coordinates of

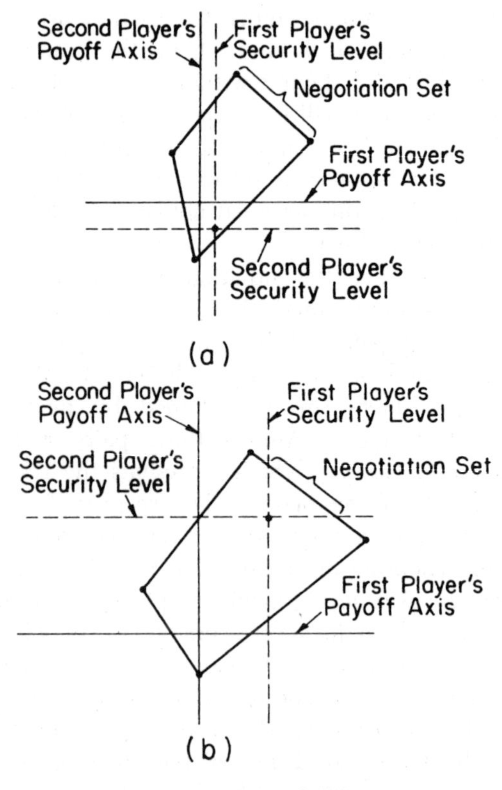

FIG. 6, (a) and (b)

the point must not represent payoffs smaller than the corresponding security levels of the two players.

We have now defined the negotiation set completely. We assume that the pair of payoffs agreed upon in a negotiation between two rational players must be represented by a point on this set.

The question before us is *which* point on the negotiation set will be agreed upon by "rational players." It is here that the notion of rationality becomes vague and somewhat mixed with other notions, such as bargaining advantage or, perhaps, equity. We can see this from the negotiation protocol reproduced above (see p. 98). Each player presents arguments which seem rational. Each points out to the other what the latter stands to lose if an agreement is not arrived at and, by implication, what an agreement ought to be worth to him. Still it is not easy to determine (objectively) the extent to which one or the other player can make his argument stick.

The fact that several theories of bargaining and arbitration have been employed in attempts to solve this problem indicates a lack of agreement among the theories' proponents. I shall here give four examples of these theories, sufficient to demonstrate the flexibility of the notions involved in the concepts of bargaining power or of equity.

First, we shall suppose that among all the points inside (or possibly on the boundary of) the payoff polygon one point is singled out as a "status quo" point. For the time being we shall not inquire which point this will be. It will merely play the part of a point of reference. The solution of the negotiated game will be sought with reference to that point. A solution will be, as we have said, a point in the negotiation set. Every point on that set, we have seen, satisfies certain properties which we feel the solution should satisfy. However, the negotiation set in general consists of many points. We should like to reduce this set to a single one. Therefore the criteria to be satisfied by that point should be more restrictive than those which define the negotiation set. Let us see what we can reasonably expect.

If the spirit of game theory is to prevail, these criteria should contain as few "psychological" overtones as possible. "Rational players," we feel, should not have any "psychology." Or, to put it another way, if they have a "psychology," it must be an extreme one: they must be perfectly rational or perfectly ruthless or (if possible) both. For if their psychologies are not extreme, then two such players might have some psychological property in different measures, which necessitates an examination of these measures. This is a psychologist's task, not a game theoretician's.

Nash's Solution of the Special Bargaining Problem

The first approach we shall examine is due to John Nash (1950). This method illustrates the mathematician's way of finding a solution to a problem when the characteristics of the solution have been specified. First, the characteristics of the solution are defined precisely. What can we expect these characteristics to be if the intrinsic (psychologically determined) bargaining abilities of the players are not to play any part in it?

1. We can expect that the solution will not depend on the way the players are labeled. Consider Games 18 and 19.

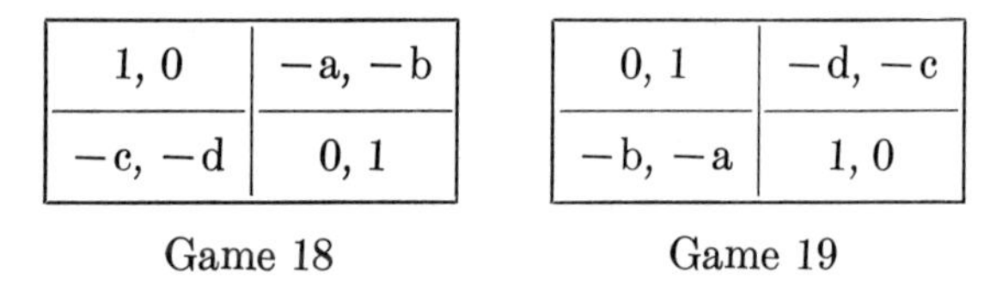

1, 0	−a, −b
−c, −d	0, 1

Game 18

0, 1	−d, −c
−b, −a	1, 0

Game 19

The only difference between them is that the roles of the two players are interchanged. Therefore the only difference in the solutions of the two games ought to be an interchange of the payoffs to the two players. In particular, if the game "looks" exactly alike to each of the players with respect to the relations between his strategies and his payoffs, then the solution should award equal amounts to each.

2. The solution ought to be in the negotiation set. We have already justified this assumption.

3. We can expect that the solution is not affected by a linear transformation of the payoffs.[15] Consider Games 20 and 21.

1, 0	−a, −b
−c, −d	0, 1

Game 20

1, 1	−a, 1 − 2b
−c, 1 − 2d	0, 3

Game 21

Game 21 is obtained from Game 20 by applying the transformation $2x + 1$ to player 2's payoffs; i.e., each of player 2's payoffs has been doubled and then increased by one unit. Suppose a solution of Game 20 prescribes the payoffs to the two players as the point on the line segment joining (1, 0) and (0, 1) and dividing this segment in proportion p:q. Then the solution of Game 21 ought to be a point on the line segment joining (1, 1) and (0, 3) and dividing that segment in the same proportion.

Note that the negotiated settlement does not really prescribe each player's "share of the joint payoff," because "joint payoff" cannot be defined in the context of utilities measured on an interval scale. The sum of the payoffs along the negotiation set is not necessarily constant, and so the "joint payoff" is not defined. What the negotiated solution prescribes is this: if the conflict is about which of several outcomes on the negotiation set is to obtain, the negotiated settlement singles out some of these outcomes and prescribes in what proportion they are to be mixed. In other words, what is prescribed is "how much each player is to have his way" in the occurrence of his preferred outcome. It is this selected set of outcomes and the proportions in which they are mixed which remain invariant when payoffs are subjected to a linear transformation.

4. Suppose the payoff polygon is enlarged so that now

additional payoff pairs become available, while the status quo point (see p. 103) remains unchanged. Then these additional payoff pairs should either contain the solution of the game or else they must not affect the solution of the old game.

This last criterion is sometimes called "independence from irrelevant alternatives." It implies that in a negotiated game, if both sides reject alternatives which are offered, the result of the negotiation should not be affected by them, provided these alternatives do not change the status quo point, i.e., the reference point of the bargaining procedure.

The problem now is to find a point in the negotiation set which satisfies these four criteria. Note that the negotiation set is a line (usually a portion of a boundary of a polygon) in the payoff space. This line can be represented by expressing Y (player 2's payoff) as a function of X (player 1's payoff). Let this function be $Y = f(X)$. Then the coordinates of the solution will be some specific pair of points $[X^*, f(X^*)]$. It was shown by J. Nash (1950) that the *only* point which satisfies the above four criteria is the point obtained by finding the maximum of the function $[X - x_0][f(X) - y_0]$ where (x_0, y_0) are the coordinates of the status quo point.

We have called this the solution of the special bargaining problem, because it is based on the assumption that the status quo point is given. There remains, however, the question of how this status quo point is to be determined. The bargaining problem with this question included we shall call the *general bargaining problem.*

In some situations the selection of the status quo points seems straightforward enough. For example, let player 1 be a buyer and player 2 a seller, and let them bargain about the price of some object to be sold by player 2 to player 1. Assume that the utility can be measured in money and that the utility scales are linear with money for both players. When the object is sold, the seller loses the utility of the object (to him) in return for the price, while the buyer loses the corresponding amount of money in return for the utility of the object (for him). Presumably a bargain will be struck if both parties gain utility as a result of their negotiations. The negotiation set then will be all possible sale prices which result in positive utility gains for both parties. It is natural to take for the status quo point the pair of utility gains associated with the outcome No Sale. Since the zero points of the utility scales of both parties can be fixed arbitrarily, this can be taken as the point (0, 0).

In the case of a game like Game 22, choice of the status quo point is not as clear.

	P_1	P_2
C_1	2, 4	−3, −2
C_2	−1, −5	4, 1

Game 22

To choose (0, 0) would be tantamount to saying that the players may decide not to play the game at all. This alternative may well be included in the formal representation of the game, as shown in Game 23.

	P_1	P_2	P_3
C_1	0, 0	0, 0	0, 0
C_2	0, 0	−2, 4	−3, −2
C_3	0, 0	−1, −5	2, 1

Game 23

Here, if either player chooses strategy 1, the outcome is (0, 0) for both, regardless of what the other does. This choice of strategy, therefore, is equivalent to a refusal to play.

But suppose it is impossible to refuse to play. What then shall be taken as the status quo point? We shall describe two possible views on this question.

Shapley's Solution

The following approach is due to L. S. Shapley (1953).

Consider the matrix of payoffs accruing to Man in Game 17 without regard to the payoffs accruing to Woman, as shown:

O_1	1	$-a$
F_1	$-c$	0

Game 24

Suppose Man solves this game as a zero-sum game. Then he has at his disposal a mixed strategy which guarantees him a certain minimum expected payoff, namely the strategy $\left(\frac{c}{1+a+c}, \frac{1+a}{1+a+c}\right)$. The guaranteed minimal payoff resulting from this strategy is $-ac/(1+a+c)$.

Now look at the game from the Woman's point of view:

O_2	F_2
0	$-b$
$-d$	1

Game 25

Woman has at her disposal the mixed strategy $\left(\frac{1+b}{1+b+d}, \frac{d}{1+b+d}\right)$, which guarantees her a minimal expected payoff of $-bd/(1+b+d)$.

Shapley's solution of the general bargaining problem amounts to taking this pair of minimal guaranteed payoffs, i.e., the security levels of the two players, as the status quo point and then solving the resulting special bargaining problem by Nash's method.

Let us see what Shapley's solution prescribes in the Battle of the Sexes.

To satisfy Nash's axioms, the following quantity must be maximized:

$$\left[X + \frac{ac}{1 + a + c}\right]\left[Y + \frac{bd}{1 + b + d}\right] \qquad (48)$$

Since along the negotiation set, $Y = 1 - X$, (48) becomes

$$\left[X + \frac{ac}{1 + a + c}\right]\left[1 - X + \frac{bd}{1 + b + d}\right] \qquad (49)$$

To maximize the quantity given by (49), we set its derivative equal to zero[16] and solve for X, namely

$$X = X^* = \frac{1}{2} + \frac{1}{2}\left[\frac{bd}{1 + b + d} - \frac{ac}{1 + a + c}\right] \qquad (50)$$

$$Y = Y^* = \frac{1}{2} + \frac{1}{2}\left[\frac{ac}{1 + a + c} - \frac{bd}{1 + b + d}\right] \qquad (51)$$

Let us see what the solution says. First, observe that if $a = b$ and $c = d$, the expressions in the brackets vanish, so that X^* and Y^* both reduce to $\frac{1}{2}$. This is, of course, exactly as it should be. If the situations of both players are exactly equal, the only equitable bargain is one which splits the difference for which the players are contending. This is accomplished by mixing the (coordinated) outcomes (O_1, O_2) and (F_1, F_2) in equal proportions.

Next note that if c and d are interchanged in (50) and also a and b, the second term on the right side of (50) changes sign and becomes the payoff due Woman. Hence Nash's first postulate is satisfied.

Next, it is clear that the payoff pair given by (50) and (51) is on the negotiation set, because both payoffs are nonnegative and add up to one (which is true only of the points on the negotiation set). Thus Nash's second postulate is satisfied.

Next, we can verify that if we multiply the payoffs of either player by a constant or add a constant to all the payoffs of either player, then the prescribed mixture of (O_1, O_2) and (F_1, F_2) will remain invariant. Thus Nash's third postulate is satisfied.

Finally, the solution depends only on the status quo point and on the negotiation set. This implies that Nash's fourth postulate is satisfied.

Now we can answer the question as to how unequal situations of the players affect the solution. We have seen that the payoffs at (O_1, O_2) and at (F_1, F_2) have been fixed by our choice of zero and unit points of the utility scales. Consequently, only a, b, c, and d can vary.

Using the method of differential calculus, we can readily see how X^* depends on each of these parameters. Differentiating X^* partially with respect to each in turn we obtain

$$\frac{\partial X^*}{\partial a} = \frac{-c(1 + c)}{(1 + a + c)^2} < 0, \tag{52}$$

$$\frac{\partial X^*}{\partial b} = \frac{d(1 + d)}{(1 + b + d)^2} > 0, \tag{53}$$

$$\frac{\partial X^*}{\partial c} = \frac{-a(1 + a)}{(1 + a + c)^2} < 0, \tag{54}$$

$$\frac{\partial X^*}{\partial d} = \frac{b(1 + b)}{(1 + b + d)^2} > 0. \tag{55}$$

The signs of the corresponding derivatives of Y^* are, of course, reversed. The meaning of inequalities (52) and (53) is clear. The greater the threat which Man can use against Woman, i.e., b, the greater is Man's share.

The greater the threat which Woman has against Man, i.e., a, the greater will be Woman's share.

The meaning of inequalities (54) and (55) is less clear. These inequalities state that the greater the loss one incurs as a consequence of misplaced altruism, the smaller is one's share in the negotiated solution. Formally it is easy to see why this is so. From (50) and (51) we see that a player's advantage is expressed as the excess of his security level over that of the other player (observe the quantities in the brackets). Therefore whatever depresses a player's security level is to his disadvantage. However, why the payoff for misplaced altruism should have an effect on the negotiated solution is not intuitively obvious. That it does have an effect is simply a consequence on the way Shapley's negotiated solution is conceived. It is derived with reference to a situation in which each player considers only his own position, essentially assuming the worst situation, namely playing against an opponent whose interests are directly opposite to his. *Such* an opponent is able to utilize every opportunity to drive one's security level down. The fact that the point, misplaced altruism, would not be normally used as bargaining leverage does not enter the derivation of Shapley's solution.

Nash's Solution of the Generalized Bargaining Problem

The essential difference between the solution proposed by Nash (1953) and that proposed by Shapley is in the choice of the status quo point. In Shapley's solution, we recall, the status quo point is determined by the players' security levels, namely by what each could get *without* bargaining, just by mixing his strategies so as to insure a certain minimum expected payoff. Reasonable as this point of reference appears, some game theoreticians feel that it does not reflect the relative *bargaining* positions of the players. Surely the players should have stronger bargaining points at their disposal than "what they can do by themselves." For instance, retreating to his own security level is not the only thing a player can do; he can also prevent the other player from getting more than *his* security level. To be sure, the payoffs associated with the threat point partially determine the security level and thus the solution. But in Shapley's solution the difference in the threat potentials can sometimes be completely obscured, as is seen in Game 26.

2, 1	−1, −2
−2, −1	1, 2

Game 26

We can verify that in Game 26 the security level of each player is zero. Since the origin of coordinates in this case lies on the perpendicular bisector of the negotiation set, the solution assigns an equal share to each player. Yet the row chooser's threat potential is larger than the column chooser's. This advantage fails to be reflected in the solution.

Nash's solution of the general bargaining problem differs from Shapley's in the way the status quo point is determined, namely not by the security levels of the players but by their choice of *threat strategies*. Roughly speaking, Nash's solution favors the player who combines a certain degree of prudence with a certain degree of brinksmanship. The exact mixture is determined by a strategic calculation, as we shall see.

A threat strategy is simply one of the available pure or mixed strategies chosen with a view of establishing a status quo point. As we have said, any pair of available strategies determines a pair of payoffs within the payoff polygon. The pair of payoffs determined by the pair of threat strategies can, in a way, be considered to be the status quo point. It represents the payoff to each player *if each carries out his threat.* Nash's solution of the general bargaining problem amounts to finding a "proper" pair of threat strategies, and so the status quo point, to which the solution of the resulting special bargaining problem is then applied.

It is therefore to each player's advantage to choose the threat strategy in such a way that the resulting status quo point will give him the greatest possible share of the joint payoff in the negotiated solution.

Let us solve the Battle of the Sexes game by this method.

Let Man's threat strategy be x. This means that Man chooses strategy O_1 with probability x (hence strategy F_1 with probability $1 - x$), the fraction x being for the time being unknown. Similarly let Woman's threat strategy be y, which means that she chooses O_2 with probability y (hence F_2 with probability $1 - y$). Then Man's expected payoff will be

$$x_0 = xy - ax(1 - y) - c(1 - x)y. \qquad (56)$$

Woman's expected payoff will be

$$y_0 = -bx(1 - y) - d(1 - x)y + (1 - x)(1 - y). \qquad (57)$$

We have taken this pair of (as yet unknown) coordinates (x_0, y_0) as the coordinates of the status quo point. Now we apply Nash's solution of the special bargaining problem. The solution of our game with respect to an arbitrary point (x_0, y_0) gives Man

$$X = 1/2 + 1/2(x_0 - y_0) \qquad (58)$$

and Woman

$$Y = 1/2 + 1/2(y_0 - x_0). \qquad (59)$$

We now write down the expression for this payoff, where x_0 is given by (56) and y_0 by (57). This gives Man

$$X^* = 1/2 + 1/2[xy + (b - a)x(1 - y) + (d - c)(1 - x)y - (1 - x)(1 - y)] \quad (60)$$

Clearly Man wishes to maximize the expression in the bracket, which represents his advantage, while Woman wishes to minimize the same expression.

Consider now the following *zero-sum* game between Man and Woman.

	W_1	W_2
M_1	1	$b - a$
M_2	$d - c$	-1

Game 27

Suppose Man in this game plays the mixed strategy $(x, 1 - x)$ and Woman the mixed strategy $(y, 1 - y)$. Then the expression in the brackets of (60) will be the expected payoff to Man, which Man would like to maximize and Woman would like to minimize. Therefore if we find these maximizing-minimizing strategies x and y, we shall have also found the x and y which maximize-minimize the expression in the brackets of (60). This pair (x, y) will therefore be the threat strategies which we are seeking.

We observe first that since the absolute value of a, b, c, and d are all less than 1, also $|a - b| < 1$. But this means that Game 27 has a saddle point, namely (M_1, W_2), because whatever the sign of $b - a$, this difference is less than 1 and greater than -1. Then the pair of strategies which constitute the solution of Game 27 are not mixed but pure strategies, namely M_1 for Man and W_2 for Woman. These can be taken as special cases of mixed strategies by setting $x = 1$, $y = 0$. Substituting these values into (60) gives Man a payoff of

$$X^* = 1/2 + (b - a)/2 \qquad (61)$$

and to Woman

$$Y^* = 1/2 + (a - b)/2. \qquad (62)$$

It follows at once that this negotiated solution does not depend on the misplaced altruism point. This circumstance pleases those who argue that the payoffs associated with misplaced altruism are irrelevant to the bargaining positions of the plays. Note, however, that our argument depends on the assumption that $|b - a| < 1$. If this is not the case, Game 27 may not have a saddle point, in which case its solution would be in terms of mixed strategies x and y. Then, if the expression in the bracket does not vanish and if $c \neq d$, the negotiated payoff to Man will depend on c and d, the payoffs associated with the misplaced altruism point. We shall not pursue the analysis of this varient.

Raiffa's Solution

A somewhat different approach to the negotiated game was proposed by H. Raiffa (1953). The nonzero-sum game, Raiffa argues, has two components, a competitive component and a cooperative component. The cooperative component reflects the fact that rational players will negotiate in such a way as to arrive at a payoff which is *jointly* the largest. After having assured for themselves the maximum joint payoff, they can proceed to compete, i.e., each will try to get the largest possible share of this joint payoff.

Now the nonzero-sum game can be turned into a zero-sum game by replacing the pairs of payoffs in each outcome with their differences. If we do so for Game 17, we arrive at the zero-sum game identical with Game 27.

Suppose now the two players play this game first, to decide what the difference between their payoffs shall be. We have already noted that Game 27 has a saddle point, namely at (M_1, W_2). Therefore its outcome is determined: Man gets b − a. Judging by this outcome of the *competitive* game, Man is "entitled" to b − a more units of payoff than Woman.

The salient feature of Raiffa's solution is that the player who gets the advantage in the outcome of the zero-sum game should keep the same numerical advantage in his share of the joint payoff. This principle seems no less reasonable than Shapley's or Nash's, but it differs in an important respect: the solution just proposed is not invariant with respect to linear transformations of the payoffs. Specifically, in the Battle of the Sexes, if we decide to double all of Man's payoffs (which simply amounts to measuring them in units one half as large as before), Nash's and Shapley's methods will lead to exactly the same respective mixtures of (O_1, O_2) and (F_1, F_2) as the game with the original payoffs. But Raiffa's method (unless it is supplemented by a normalization, to be presently described) will lead, in general, to a different mixture if the payoffs undergo a linear transformation. This defect (if it is a defect) is removed by establishing a standard utility scale before the procedure is applied. Namely, the zero and the unit points of both players are fixed in a particular way instead of being chosen arbitrarily, as they may be in either Nash's or Shapley's treatment of the bargaining problem.

The payoff scale proposed by Raiffa is the following. Let the worst outcome in each player's set of outcomes be associated with payoff 0, and the best outcome with payoff 1. These two points of the utility scale being fixed, the other payoffs are thereby also determined by the ratios of the intervals between them. Since this procedure is a linear transformation, any two sets of payoffs which are linear transformations of each other will yield the same set of payoffs when normalized in the way described.[17] In this way, Raiffa's solution remains invariant if the originally given payoffs are changed by a linear transformation. But the *procedure* described above (namely, solving the competitive game and using the solution to split the joint payoff) cannot be applied to a set of payoffs on an arbitrarily chosen scale. First the scale must be transformed to the standard scale. Only then can the procedure be utilized.

When this normalization is applied to the Battle of the Sexes, we see that a and b must be assigned value 0, while 1 remains 1. Then, prescribing interval ratios, we must transform c into $(a - c)/(1 + a)$, d into $(b - d)/(1 + b)$, Man's 0 into $a/(1 + a)$, and Woman's 0 into $b/(1 + b)$.

The resulting game becomes

	O_2	F_2
O_1	$1, \dfrac{b}{b+1}$	$0, 0$
F_1	$\dfrac{a-c}{1+a}, \dfrac{b-d}{1+b}$	$\dfrac{a}{a+1}, 1$

Game 28

The zero-sum game derived from the differences of the payoffs becomes

	W_1	W_2
M_1	$\dfrac{1}{b+1}$	0
M_2	$\dfrac{a-c}{1+a} - \dfrac{b-d}{1+b}$	$-\dfrac{1}{a+1}$

Game 29

Game 29 has a saddle point at (M_1, W_2) where the payoffs of both players are zero. So it appears that neither player has a numerical advantage in the zero-sum game normalized by Raiffa's scheme. The equation of the negotiation set line now becomes

$$\left(Y - \frac{b}{1+b}\right) = -\frac{b(1+a)}{a(1+b)}(X-1). \qquad (63)$$

Setting $X = Y$ (since neither has a numerical advantage) and solving for X, we obtain

$$X^* = \frac{ab + b(1+a)}{a(1+b) + b(1+a)}. \qquad (64)$$

Reconverting into our original units, we get the share due Man and Woman respectively:

$$X^* = \frac{b(1+a)}{a(1+b) + b(1+a)};$$
$$Y^* = \frac{a(1+b)}{a(1+b) + b(1+a)}, \qquad (65)$$

which is Raiffa's negotiated solution for the Battle of the Sexes.

We note that the solution behaves properly in several respects. As in Nash's solution, the shares depend only on the available threats and not on the misplaced altruism point. Note, however, that as in the preceding case, this is partially a consequence of our restrictions on the payoffs. For example, had we not assumed that the payoffs at the threat point were the worst for both players, the normalized payoffs for that outcome would not both have been zero, and so the solution of the zero-sum game would have been different. The final solution may then have involved the misplaced altruism point.

Finally, we can verify by methods already described that Man's share increases with his threat capability and decreases with Woman's threat capability, as we should expect.

Braithwaite's Solution

R. Braithwaite (1955) proposed a method of arbitration similar to Raiffa's. Again the players decide on a division of the joint payoff by solving the zero-sum game derived from the differences of their payoffs. Braithwaite's method differs from Raiffa's in the choice of normalization. As for Raiffa, for Braithwaite, "equity" is essentially the problem of choosing appropriate units in which to measure each player's payoff.

In a nonzero-sum game there is a distinction between a *maximin* strategy and a *minimax* strategy. The maximin strategy is one which assures a player the minimal guaranteed payoff (the security level). The minimax strategy is one which keeps the *other* player's payoff to his security level.[18] It turns out that if one player shifts from his maximin strategy to his minimax strategy, while the other player sticks to his maximin strategy, the former player's payoff increases. Braithwaite's normalization scheme is a choice of payoff units which insures equal increases of this sort to each player. Note that the notion of "threat" is implicit in this procedure, because a shift to the minimax is essentially a method of keeping the opponent's payoff to its minimum.

In our Battle of the Sexes game, Man's minimax strategy is $\left(\frac{1+d}{1+d+b}, \frac{b}{1+d+b}\right)$; Woman's minimax strategy is $\left(\frac{a}{1+a+c}, \frac{1+c}{1+a+c}\right)$. The calculation of the payoff increases associated with the shifts from maximin to minimax are straightforward but tedious. These shifts will be equalized if Man's payoffs are multiplied by the factor $(1 + b + d)$, while Woman's are multiplied by the factor $(1 + a + c)$. When this is done, the zero-sum game, in which the payoffs are the differences of Man's and Woman's payoffs gives Man an advantage of

$b(1+c) - a(1+d)$. This "advantage" can, of course, be negative if $a(1+d) > b(1+c)$, which happens if a or d are sufficiently large compared to b and c. The final solution (converted to the original payoff units) gives Man and Woman respectively

$$X^* = \frac{1+b+c+bc-ad}{2+a+b+c+d}, \qquad (66)$$

$$Y^* = \frac{1+a+d+ad-bc}{2+a+b+c+d}. \qquad (67)$$

Comparison of the Four Solutions

It is clear that if $a = b$, $c = d$, all four solutions prescribe an equal mixture between opera and prize fight. We can therefore compare the four solutions in terms of the excess over $\frac{1}{2}$ which each awards to Man. This "excess" can, of course, be either positive or negative, depending on the relations among the parameters a, b, c, and d. The excess to Man in each of the four methods is as follows:

Shapley: $$1/2\left[\frac{bd}{1+b+d} - \frac{ac}{1+a+c}\right] \qquad (68)$$

Nash: $$1/2\,[b-a] \qquad (69)$$

Raiffa: $$1/2\left[\frac{b-a}{a(1+b)+b(1+a)}\right] \qquad (70)$$

Braithwaite:

$$1/2\left[\frac{(b-a)+(c-d)+2(bc-ad)}{2+a+b+c+d}\right] \qquad (71)$$

The simplest form of settlement is that of Nash. Man (or Woman) gets his (or her) way to the extent that his (or her) threat potential exceeds hers (or his).

Raiffa's settlement also depends only on the threat potentials, but in a more complicated way. We see also that when the threat potentials are small, the excess given to the player with the larger threat potential is larger in Raiffa's settlement than in Nash's; but when the threat potentials are large, the opposite is true.

Shapley's and Braithwaite's settlements both involve the misplaced altruism points as well as the threat potentials. There is a fundamental difference between these two settlements, namely in the role of the misplaced altruism payoff. In Shapley's settlement a large (negative) misplaced altruism payoff is detrimental to the player who has it; in Braithwaite's settlement, on the contrary, it benefits the player who has it. We have already explained the role of misplaced altruism in Shapley's solution: it depresses the security level and thus the negotiated payoff. How can we explain the opposite result in Braithwaite's settlement? Observe that a large negative payoff for misplaced altruism carries the "message" that the associated outcome will be strongly *avoided*, which also means that concessions will be hard to get from the player threatened with this outcome. This puts the player in a favorable bargaining position.

Thus we see that from the point of view of final settlements, Shapley's and Braithwaite's solutions stand at opposite poles (because of the opposite roles played in them by the misplaced altruism payoffs). Those of Nash and Raiffa fall between (because misplaced altruism plays no part in these solutions).

From the point of view of the underlying rationale, however, Shapley's and Nash's procedures belong to one type while Raiffa's and Braithwaite's belong to another. For Shapley's and Nash's schemes reduce to Nash's solution of the special bargaining problem after the status quo point has been determined. The two methods differ only in the way the status quo point is determined. On the other hand, both Raiffa's and Braithwaite's methods are based on the idea of breaking up the original game into a competitive and a cooperative phase. These two methods differ only in the way the utility scales of the two players are to be normalized.[19]

It must be borne in mind that the above analysis was carried out only in one type of game used as an illustration, namely a Battle of the Sexes game with certain restrictions on the payoff parameters. To what extent the results are generalizable to other types of games is still an open question.

9. Nonnegotiable Games

Consider the following game

	P_1	P_1
C_1	$-1, -1$	$1, 1$
C_2	$2, -2$	$-2, 2$

Game 30

Suppose we treat it as a negotiable game and solve it by Shapley's method. Broken up into separate games, Game 30 becomes

C_1	-1	1
C_2	2	-2

Game 31
(Castor's)

P_1	P_2
-1	1
-2	2

Game 32
(Pollux')

Castor's maximin strategy is $(\frac{2}{3}, \frac{1}{3})$. His security level is 0. Pollux' maximin strategy is P_2 since it is the dominating strategy, and his security level is 1. The payoff polygon is shown in Figure 7.

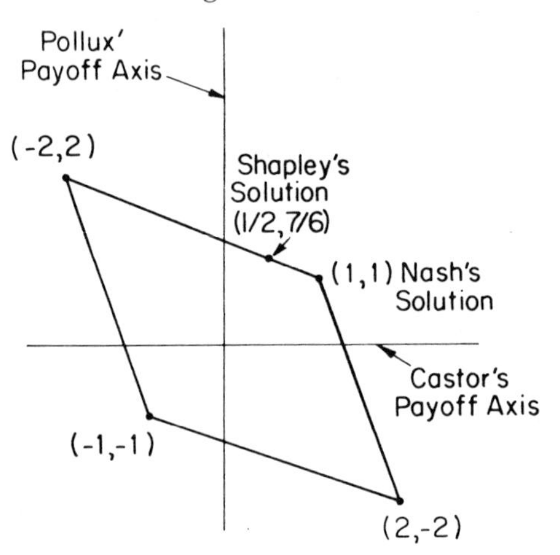

FIG. 7. The payoff polygon of Game 30.

The negotiation set is the line segment connecting (0, 4/3) to (1, 1). The equation of this line is $Y = 4/3 - X/3$. Accordingly we maximize $X(4/3 - X/3 - 1)$ and obtain

$$X^* = 1/2, \; Y^* = 7/6. \tag{72}$$

The solution prescribes a mixture of the outcomes (1, 1) and (−2, 2) in proportion (5/6, 1/6).

Let us now solve the game by Nash's method. Formal application of the method leads to C_1 as Castor's best "threat" strategy and to P_2 as Pollux' best "threat" strategy.[20] Hence (1, 1) is the status quo point.

We maximize

$$(X - 1)(1/3 - X/3) \tag{73}$$

and obtain

$$X^* = Y^* = 1. \tag{74}$$

Which is the more reasonable solution. Shapley's or Nash's? Pollux gets more in Shapley's solution. Does he "deserve" to get more? To see whether he does, let us see what would happen if the two players made their decisions *without negotiation, but with full knowledge of each other's payoffs.*

Note that this question is not identical to the question concerning the players' security levels. Security levels are determined *in ignorance* of the other's payoffs or, to put it in another way, by assuming the worst about the other player, namely that his interests are diametrically opposed to one's own. If, however, the payoffs of the other are available, decisions can be made on another basis. In particular, Pollux' decision in Game 30 is obvious. Since P_2 dominates P_1, he must choose P_2. Castor, seeing Pollux' payoffs, will also conclude that Pollux will choose P_2, since P_2 dominates P_1. Castor's best answer to P_2 is C_1. Accordingly, in the absence of negotiation, the outcome (C_1, P_2) will obtain, which gives each player 1 unit. This is also the outcome recommended by Nash.

Because of the decisive role played by the dominating strategy, it is difficult to disagree with this verdict. The only argument in favor of Shapley's solution is that Pollux deserves some bonus because his security level is higher than that of Castor's. But this argument is hard to accept if both players are fully aware of the game they are playing.

Nash's solution can be defended on still another ground. The outcome (C_1, P_2) is a so-called equilibrium point. This means that neither player is willing to move away from this outcome, once it obtains. For if Castor should move away, he can move only to (C_2, P_2) which is worse for him than (C_1, P_2); while if Pollux is to move away, he can only move to (C_1, P_1) which is also worse for him. An equilibrium outcome, in other words, is one from which neither player is motivated to move away (alone), for if one does he can only do worse or, at least, no better.

It would seem, then, that the concept of equilibrium is a useful one in the treatment of the nonzero-sum game, especially since it enables us (possibly) to solve games without appeal to negotiation. Recall also that the solutions of zero-sum games are also equilibria and that those games are all nonnegotiable.

A nonzero-sum game can have more than one equilibrium. Consider the following game:

	P_1	P_2	P_3
C_1	3, 1	0, 0	0, 0
C_2	0, 0	2, 2	0, 0
C_3	0, 0	0, 0	1, 3

Game 33

Here (C_1, P_1), (C_2, P_2), and (C_3, P_3) are all equilibria, since neither player is motivated to move away from either of these outcomes. The two players, however, have opposite orders of preference for the equilibrium outcomes. Castor, evidently, likes (C_1, P_1) best, while Pollux likes (C_3, P_3) best.

Because of the symmetry of this game, we know that each of the four methods of solving negotiated games will lead to an equal payoff to each player, i.e., to the outcome (C_2, P_2). It has also been argued, for example by T. C. Schelling (1960), that players will arrive at the same outcome even without negotiation. In fact, it might seem that the outcome (C_2, P_2) is even more likely without negotiation than as a result of negotiation. For if we forget for the moment the cardinal principle of game theory, namely that players are psychologically indistinguishable, we might conjecture that the "stronger" negotiator (say, Castor) might get his way by threatening to stick with C_1 regardless of what Pollux does. Pollux has no recourse against this threat but to yield (unless he wants to punish himself as well as Castor), i.e., choose P_1.[21] Of course, Pollux can insist on P_3 if he is the "stronger" negotiator.

Note that if negotiation is impossible this sort of blackmail cannot take place. It does Castor no good to play C_1 if Pollux does not know of his intent. Threats which cannot be transmitted are powerless. Players who cannot communicate must depend entirely on guesses about what the other is going to do. In some cases such guesses can be defended on rational grounds. For instance, in the case of Game 30, Castor's best guess is certainly that Pollux will choose P_2. Can a reasonable guess be made in the case of Game 33? Schelling (1960) argues that it can, namely, that each player can depend on the other's choosing strategy 2. Schelling calls strategy 2 a *prominent* strategy. First, it is an equilibrium; second, among the three equilibrium strategies in this game, it is the "prominent" one, singled out because of its symmetry. Neither of the other two equilibrium strategies (C_1, P_1) and (C_3, P_3) can claim uniqueness, because each of them is a mirror image of the other. While (C_1, P_1) favors Castor, and (C_3, P_3) favors Pollux (C_2, P_2) is unique: it favors neither player.[22]

Schelling argues that wherever it is of interest to both players to *coordinate* their choices (as it obviously is in Game 33), they will (or should) seek a pair of strategies which somehow stand out, so that even in the absence of negotiation each can depend on the other's good sense to choose the strategy which offers the opportunity for a *tacit* agreement.

Game 33 can be generalized. Suppose Castor and Pollux are to name independently some fraction of a dollar. If the sum of the fractions which they name does not exceed a dollar, each will get the fraction he names. If, however, the sum does exceed a dollar, neither will get anything. In this game, any pair of fractions (x, y) such that $x + y = 1$ is an equilibrium. For each player hurts himself by claiming either less or more. If, for example, the first player chooses $x' < x$, while the second chooses y, the first player will get $x' < x$. If he chooses $x'' > x$, while the other chooses y, he will get nothing. Of all these equilibria, however, one is "prominent," namely $(\frac{1}{2}, \frac{1}{2})$. It seems reasonable, therefore, that in the absence of negotiation, each player should choose $\frac{1}{2}$. If the game is negotiated, each of the four methods described in the previous chapter will also prescribe $\frac{1}{2}$ to each player.

J. Nash has shown (1951) that every nonzero-sum game has at least one equilibrium point. Coupled with the notion of prominence, the notion of equilibrium therefore seems to extend the theory of nonzero-sum games to the nonnegotiable case. Namely, if a game has only one equilibrium, then the corresponding pair of strategies can be prescribed in virtue of the properties of the equilibrium. If there are several equilibria and one happens to be "prominent," then the pair of strategies corresponding to the prominent equilibrium might be a reasonable solution of the nonnegotiable game.

While this extension of the theory can be defended in many cases, there are situations in which it leads to paradoxes, as we shall now show.

Consider the following game.[23]

	C_2	D_2
C_1	5, 5	−10, 10
D_1	10, −10	−5, −5

Game 34, Prisoner's Dilemma

[We have changed our notation of the strategies, for reasons which will become clear. Here the players are labeled 1 and 2, and each has a choice of a C strategy or a D strategy.]

If the game is negotiable, each of the methods described in the previous chapter leads to the same solution: the outcome (C_1, C_2).

Suppose, however, the game is nonnegotiable or, which is the same thing, suppose the agreement is not binding. Then the situation looks very different. For in the absence of agreement it is obviously in the interest of each player to choose the D strategy. If an *unenforceable* agreement has been made to choose the C strategy, it is in the interest of each player to break the agreement. Moreover, it is in the interest of each player to break the agreement *regardless of whether the other keeps it or not.* If the other keeps it (chooses C) the defector can get the biggest payoff by choosing D. If one player breaks the agreement (chooses D) then there is all the more reason for the other to do so, since otherwise he is left holding the bag. Not only is a constant temptation to break the agreement pressing on each player, but also the *knowledge* that the other player is tempted makes the defection practically compelling.

Formally speaking, strategy D dominates strategy C for both players (see p. 54). In order to make the agreement to choose C stick, some incentives must be provided. In real life the most common incentives for keeping agreements are the sanctions which are imposed when agreements are broken. These may be in the form of legally imposed penalties or loss of credit, prestige, or honor. At any rate, if breaking the agreement is viewed as a move in a game, then payoffs must be assigned to outcomes associated with this move. Sanctions are payoffs with negative utility. If these negative utilities are added to the original payoffs the resulting payoffs are different, and one is playing a different game.

For this reason it has been argued by some game theoreticians, particularly Nash, that game theory should be developed primarily in the context of the nonnegotiated game where questions related to agreements need not be raised. If negotiations are provided for, they should simply be treated as moves in another game composed of the original one plus the negotiations. One could, of course, introduce a "higher order negotiation" dealing with the new game. But this process has no end. It is an open question whether one should stop with a negotiated game where agreements are absolutely binding or with a nonnegotiated game which results when the negotiations are reduced to sequences of moves.

At any rate, it is certainly in order to analyze a nonnegotiable game, because in real life negotiations are often impossible or fruitless, and also because frequently agreements cannot be enforced.

If no binding agreement can be effected, the mutually advantageous choice (C,C) is impossible to rationalize by appeal to self-interest. By definition, a "rational player" looks out for his own interest only. On the one hand, this means that the rational player is not malicious—that is, he will not be motivated to make choices simply to make the other lose (if he himself gains nothing in the process). On the other hand, solidarity is utterly foreign to him. He does not have any concept of collective interest. In comparing two courses of action, he compares only the payoffs, or the expected payoffs, accruing to him personally. For this reason, the rational player in the absence of negotiation or binding agreements cannot be induced to play C in the game we are discussing. Whatever the other does, it is to his advantage to play D. It is clear, however, that if both reason this way (and both do, if they are "rational" in the sense of game theory) both are worse off than if they were guided by their joint interest to play C. Note also that the outcome (D_1, D_2) is the *only* equilibrium in this game.[24]

We have then a genuine bifurcation of the notion of

rationality into that of individual rationality and that of collective rationality, each of which prescribes a different strategy to both players.

One might think that the situation is different if the game is played not once, but many times in succession, the result of each play being announced to both players. For in that case a player might hesitate to play D for fear of retaliation by his opponent on successive plays. On the other hand it seems sensible to play C as a way of communicating to the opposing player that one is ready to cooperate if he will. If he goes along, then a *tacit* agreement might be reached to continue to cooperate (i.e., choose C). No external sanctions need to be applied for breaking this tacit agreement. Each player is prevented from defecting to D by the knowledge that the other, in order to save himself from the worst outcome associated with unilateral cooperation, will be forced to retaliate by also playing D on successive plays. But the choice (D,D) is punishing to both. It therefore seems sensible to stick to (C,C) at least until the very last play of the game, after which no more retaliation is possible.

This last qualification, exempting the last play from the tacit agreement, proves to be the undoing of the entire argument. For, suppose the game is to be played some known number of times, say one hundred times in succession. Whatever is the prudent policy to follow during the first ninety-nine plays, it seems that on the one hundredth play it is of advantage to play D. This is so regardless of whether the other cooperates to the very end or whether he also decides to take advantage of the impunity conferred by the fact that the one hundredth play is the last. Both players, therefore, know that on the last play the outcome will be (D,D), and that they cannot prevent it during the course of the other plays. On the other hand, the last outcome cannot influence what happens earlier. Therefore the last outcome must simply be written off: it will be *unconditionally* (D,D). But if the one hundredth outcome is known and if nothing can be done about it, then the ninety-ninth outcome becomes effectively the last, and the same reasoning applies to it. Therefore the ninety-ninth outcome must also be unconditionally (D,D). In this way the entire system of tacit collusion which appeared to make strategic sense collapses like a row of dominoes, and we arrive at the bizarre conclusion that two players, pursuing their individual interests, ought to play D all the one hundred times in succession, even though this gives each of them a loss of five hundred units, whereas had they played C one hundred times in succession, they would each have won five hundred units.

We feel that even though the choice of D in a single play of the game may seem justified (in the absence of a possibility to agree on C), the choice of D one hundred times in succession is much more difficult to defend when an opportunity exists to establish a tacit agreement, enforceable at least by "deterrence." Why this is so can be seen more clearly in a formal analysis of a game matrix.

For simplicity we shall suppose that the game is played just twice. But now we shall view this succession of two plays as a single play of a game with two moves. Each move will be a simultaneous choice between C and D by the two players. The result of the first move will be known to both before the second move is made.

We can now list all the *strategies available* to each of the players in this two-move game. Recall that the choice of a strategy by each player completely determines the course of the game. The eight strategies open to each of the players are the following:

1. On the first move choose C; on the second move choose C regardless of what the other chose on the first move.

2. On the first move choose C; on the second move choose what the other chose on the first move.

3. On the first move choose C; on the second choose the opposite of what the other chose on the first move.

4. On the first move choose C; on the second choose D regardless of what the other chose on the first move.

5. On the first move choose D; on the second choose C regardless of what the other chose on the first move.

6. On the first move choose D; on the second choose what the other chose on the first move.

7. On the first move choose D; on the second choose the opposite of what the other chose on the first move.

8. On the first move choose D; on the second choose D regardless of what the other chose on the first move.

We are now in a position to display the game in matrix form. The outcomes resulting from each of the paired strategy choices are shown in Game 35.

Let us now compare Game 35 with Game 34. We see that in Game 34, D is a dominating strategy for each player. However, the larger Game 35 has no strategy which dominates all the others. In particular the unconditionally uncooperative strategy 8, while it is the best answer to the other's choice of strategy 8, is not best against all the other's choices. For example, strategy 8 is not the best answer against the other's strategy 2. Against this strategy, strategies 3 or 4 are best.

If, therefore, there were some way in which player 1 could *convince* player 2 that he would in fact play strategy 2, then player 2 would be serving his own interest by choosing strategy 4, not strategy 8. As a result player 1 would get −5 and player 2 would get 15. Player 2 would get the best of it, but both would be better off than they would have been with strategy 8, which gives each player −10.

The question is whether there is a way for player 1 to let player 2 know (and to convince him) that he will

	1	2	3	4	5	6	7	8
1	10, 10	10, 10	−5, 15	−5, 15	−5, 15	−5, 15	−20, 20	−20, 20
2	10, 10	10, 10	−5, 15	−5, 15	0, 0	0, 0	−15, 5	−15, 5
3	15, −5	15, −5	0, 0	0, 0	−5, 15	−5, 15	−20, 20	−20, 20
4	15, −5	15, −5	0, 0	0, 0	0, 0	0, 0	−15, 5	−15, 5
5	15, −5	0, 0	15, −5	0, 0	0, 0	−15, 5	0, 0	−15, 5
6	15, −5	0, 0	15, −5	0, 0	5, −15	−10, 10	5, −15	−10, −10
7	20, −20	5, −15	20, −20	5, −15	0, 0	−15, 5	0, 0	−15, 5
8	20, −20	5, −15	20, −20	5, −15	5, −15	−10, −10	5, −15	−10, −10

Game 35

play strategy 2.[25] Let us see whether player 2 can *deduce* player 1's intentions.

If he could be sure of one of two things, player 2 might well ascribe to player 1 an intention to play strategy 2: either (1) player 1 will choose strategy 2 "on commonsense grounds" (i.e., without completing the strategic analysis); or (2) player 1, having completed the strategic analysis will act "ethically" rather than strategically.

The "commonsense grounds" for choosing strategy 2 might be somewhat of this sort: if we cooperate (choose strategy 1) we will both be better off than if we do not (say choose strategy 8). I do not know whether he will cooperate, but I shall see. If he cooperates on the first move, then he probably has the same idea that I have. If he does not, I was wrong about him and so I shall resort to D.

This "commonsense" plan fails to take into account the other's "best reply" to strategy 2. The "best reply" is strategy 4, i.e., feinting cooperation on the first move in order to induce the other's cooperation in the second (which strategy 2 promises) so as to take advantage of this cooperation by defecting on the second move.

Acting "ethically" rather than strategically means *ignoring* the possibility of the other's taking advantage of one's good intentions; it means carrying out these intentions anyway. In other words, even if player 1 knows that player 2's best answer to strategy 2 is strategy 4, he may nevertheless stick with strategy 2 on the grounds that the resulting payoff of −5 is still better than −10, which each will get if each *follows the logic of strategy to the bitter end and acts on it.* This logic is epitomized in the following considerations. "If he could trust me (not to budge from my intention to make my cooperative response as the second move contingent on his cooperation on the first), it would be to his best interest to play cooperatively on the first move. But he knows that it is to my advantage to double-cross him on the second move. Therefore, he must assume that I shall do so, and so he must decide to double-cross me on the first move. But he knows that I have already deduced this decision of his and so have decided *not* to play cooperatively even on the first move. Therefore he will not either."

The only way to get out of this vicious circle is to apply some ironclad guarantee that this line of reasoning ("he may, therefore I ought, therefore he will, therefore I must choose the defecting strategy") will be nipped in the bud. But if the possibility of guarantees are considered, one might as well admit a guaranteed agreement between the players, and so we are back in the context of a negotiated game!

In view of the situation just described, can we speak of a general game-theoretical solution of nonnegotiated games? As we have seen, the game theorist is not particularly concerned with whether the outcomes which he calls "solutions" are intuitively acceptable, however one chooses to define rationality. Indeed, it turns out that no single definition of rationality suffices. If one admits several definitions, it is not surprising that a game-theoretical solution satisfies some criteria of rationality and not others. The mathematician or the game theorist is concerned rather with whether certain outcomes of a game can be singled out as somehow special. In the case of the two-person zero-sum game, as we have seen, such special outcomes (when each player plays a minimax strategy) satisfy intuitively acceptable criteria of a solution. Neither player can individually improve his payoff if both choose a minimax strategy. Since, moreover, the players of a zero-sum game cannot jointly improve their payoffs, the minimax appears as a natural solution of the two-person zero-sum game.

In the case of the nonzero-sum game, it is not clear how in the absence of an enforceable agreement or, at least, of negotiation, the two players can improve their payoffs jointly. On the other hand, there are always some outcomes of these games which result because neither player can improve his own payoff if the other sticks to the strategy leading to the outcome. It seems natural, therefore, to view these outcomes as solutions of non-zero-sum nonnegotiable games. These are the equilibria, defined above.

The role of the equilibrium in the theory of the non-negotiable game is controversial. It is tempting to view the equilibrium as the natural extension of the minimax solution of the zero-sum game because of its analogous properties. However there are serious difficulties for accepting one of the equilibria of a nonnegotiable game as "the" solution. We have already examined some of these difficulties in the context of Prisoner's Dilemma. Namely, in a "supergame" consisting of a hundred (or a million) plays of Prisoner's Dilemma, in which the outcome of each play is announced, the equilibrium strategy, of both players is the totally uncooperative strategy, i.e., the choice of D all one hundred (or million) times. This result appears absurd, no matter how impeccable the logic which leads to it. But this discrepancy between strategic logic and common sense does not exhaust the difficulties with the equilibrium solution.

Consider the following game

	C_2	D_2
C_1	1, 1	−2, 2
D_1	2, −2	−5, −5

Game 36, Chicken

This game resembles Prisoner's Dilemma (Game 34) in that both players are tempted to defect from the co-operative outcome (C_1, C_2) and both are punished in the outcome (D_1, D_2) if both defect. Chicken differs, however, from Prisoner's Dilemma in that the D strategies do not *dominate* the C strategies. The D strategy is each player's best answer to the other's C strategy, but the C strategy is the best answer to the D strategy.

Next we note that unlike Prisoner's Dilemma, which has a simple equilibrium at (D_1, D_2), Chicken has at least two equilibria, namely at (C_1, D_2) and at (C_2, D_1). Of these, player 1 prefers (C_2, D_1), while player 2 prefers (C_1, D_2). In the absence of negotiation each must choose his strategy independently of the other. Suppose each player chooses the strategy which contains his own preferred equilibrium. The resulting outcome is (D_1, D_2) which is the worst for both. Suppose, on the other hand, each chooses the strategy which contains the other's preferred equilibrium. The resulting outcome is (C_1, C_2) which is intuitively satisfactory, but is not an equilibrium.

How shall rational players choose? Let us see whether there may be other equilibria. Suppose player 1 plays the mixture $(x, 1-x)$ and player 2 plays $(y, 1-y)$. Then the respective expected payoffs will be

$$G_1 = xy - 2x(1-y) + 2(1-x)y - 5(1-x)(1-y) = x(3-4y) + 7y - 5; \quad (75)$$

$$G_2 = xy + 2x(1-y) - 2(1-x)y - 5(1-x)(1-y) = y(3-4x) + 7x - 5. \quad (76)$$

Now if $y < \frac{3}{4}$, then the coefficient of x in (75) is positive and player 1 can get more by increasing x. He gets most by setting $x = 1$, i.e., playing the pure C_1 strategy. Player 2's best answer to C_1 is D_2, and we arrive at (C_1, D_2), an equilibrium point. Similarly if $y > \frac{3}{4}$, the coefficient of x in (75) is negative and player 1 can get more by decreasing x. He gets most by setting $x = 0$, i.e., playing the pure D_1 strategy. Player 2's best answer to D_1 is C_2, and so we arrive at the other equilibrium point. Examining (76), we find that player 2 faces the same situation depending whether $x < \frac{3}{4}$ or $x > \frac{3}{4}$.

Suppose, however, $x = y = \frac{3}{4}$. Then the coefficient of x in (75) vanishes, and so does the coefficient of y in (76). Now neither player can affect his own expected payoff by choosing a mixed strategy because his payoff becomes independent of the mixed strategy which *he* controls, i.e., x in the case of player 1 and y in the case of player 2. Therefore the strategy mixture ($\frac{3}{4}$, $\frac{1}{4}$) chosen by both players represents an equilibrium of sorts. It gives each player $\frac{1}{4}$, which is above his security level (which is -2 in this game) but not as much as 1, which is what each gets at (C_1, C_2).

The advocates of the equilibrium solution would argue that $\frac{1}{4}$ is the most the players can hope to get in the absence of negotiation, because the outcome (C_1, C_2) is "inaccessible" if the two cannot coordinate their strategies (Harsanyi, 1962). But whether it is "inaccessible" or not depends on how one defines rational choice.

Suppose, for example, we identify rationality with prudence. The prudent strategy is the one which guarantees one's security level. In Game 35, player 1's prudent strategy is C_1 (because it contains his security level -2), and player 2's prudent strategy is C_2. The outcome is (C_1, C_2) which gives each player $+1$ instead of $+\frac{1}{4}$, which each receives at equilibrium. Against this one can argue that ($\frac{3}{4}$, $\frac{1}{4}$) is a better answer to C_1 than is C_2, for the mixture gives player 2 expected payoff of 5/4 instead of 1. But if one makes this argument, why not go all the way and argue that D_2 is a still better answer to C_1, because it gives player 2 a payoff of 2? The reply is that one cannot expect his opponent to play the pure C_1 strategy if he is "rational." And this again raises the question of what "rational" is.

J. C. Harsanyi (1962) argues that the concept of rationality should include the "mutual perception of each other's rationality." But whether one perceives the other as "rational" depends on how one *defines* rationality. Therefore Harsanyi's criterion is circular. So are other criteria. Take prudence. If I call prudence rational, and expect the other to be rational, I expect him to be prudent. Therefore I can expect him to play C. Can I *on the basis of this assumption* rationalize an attempt to get more by playing D? Not without abandoning either my definition of rationality or my claim to rationality!

Similarly the equilibrium strategy $(\frac{3}{4}, \frac{1}{4})$ depends on the assumption that the other is like me. So the only reason for him to choose the mixture is the assumption that I shall play it and moreover shall expect him to play it. The mixture is the only "prominent" one among the three equilibria. But its prominence is due only to its mathematical symmetry, not to any advantage it confers on the players.

If one chooses "symmetry" as a basis for coordinating strategies *without* demanding that the solution be an equilibrium, then the choice in Game 35 is clearly among the following three pairs: (C_1, C_2), the pair of mixtures $(\frac{3}{4}, \frac{1}{4})$, and (D_1, D_2). Of these three (C_1, C_2) benefits both players most, and I, for one, fail to see any good reason for refusing to name it as the "rational" solution of the nonnegotiated game.

To see the situation from still another point of view let us see what a player can expect from the symmetrical equilibrium mixture $(\frac{3}{4}, \frac{1}{4})$. Setting $x = \frac{3}{4}$, player 1 examines his expected gain

$$\begin{aligned} G_1 &= 3/4(3 - 4y) + 7y - 5 \\ &= -\frac{11}{4} + 4y. \end{aligned} \tag{77}$$

We see that player 1 with his equilibrium mixture is entirely at the mercy of player 2. By setting $y = 0$, player 2 can make player 1 get $-\frac{11}{4}$, which is below player 1's security level, while player 2's payoff remains at $\frac{1}{4}$. "But why should player 2 do this?" counters the proponent of the equilibrium solution. When player 1 plays the equilibrium mixture, player 2 cannot affect his *own* payoff at all and so has no incentive to depress player 1's payoff. But there is such an incentive if player 2 anticipates player 1's *response* to setting $y = 0$. Player 1's best response is to set $x = 1$, which gives player 2 his biggest payoff (2).

To this the advocate of the equilibrium replies:

"Oh, but player 1 can also do what player 2 is contemplating. Since both players are rational, each must refrain from making any 'asymmetrical assumption' which promises him an advantage." So goes the argument involving the "mutual perception of rationality." But once one invokes such an argument, why not invoke it to rationalize (C_1, C_2) which gives both players more than the equilibrium and which can be, moreover, supported by extending the security level principle to this game?

Also if "mutual perception of rationality" is a valid criterion of rationality, can it not be extended even to Prisoner's Dilemma? The argument for choosing C in that game might go something like this (player 1 speaking).

"The best outcome for both of us is (C, C). However, if player 2 assumes that I shall choose C, he may well play D to win the largest payoff. To protect myself I will also play D. But this makes for a loss for both of us. Two rational players certainly deserve the outcome (C, C). I am rational and by the fundamental postulate of game theory, I must assume that player 2 is also rational. If I have come to the conclusion that C is the rational choice, he too must have come to the same conclusion. Now knowing that he will play C, what shall I play? Shall *I* not play D to get the greatest payoff? But if I have come to this conclusion, he has also probably done so. Again we end up with (D, D). To insure that he does not come to the conclusion that he should play D, I better avoid it also. For if I avoid it and am rational, he too will avoid it if he is rational. On the other hand, if rationality prescribes D, then it must also prescribe D for him. At any rate because of the symmetry of the game, rationality must prescribe *the same choice to both.* But if both choose the same, then (C, C) and (D, D) are the only possible outcomes. Of these (C, C) is clearly the better. Therefore I should choose C."

It is perhaps remarkable that while some game theoreticians might accept the conclusion that C is the rational choice in the game of Chicken, they usually do not accept the conclusion that C is the rational choice in Prisoner's Dilemma. Perhaps the reason is that in Chicken the choice of C is *also* the choice of one's own maximin strategy while in Prisoner's Dilemma the choice of C is not the choice of one's maximin strategy. In other words the idea of the maximin dominates the concept of rationality in some game theorists' analysis of nonzero-sum games as well as of zero-sum games.[26]

The foregoing examples lead us to the following conclusion. Either the concept of rationality is not well defined in the context of the nonnegotiable nonzero-sum game; or if the definition of rationality in the context of the zero-sum game is applied to the "solution" of some nonzero-sum games, the results are paradoxical. At any rate neither the maximin nor the equilibrium is a satisfactory concept as a basis for the solution of all nonzero-sum games. We have seen that in Prisoner's Dilemma, the choice of maximin leads to an equilibrium outcome which is bad for both players, while in Chicken, if each player chooses the strategy which contains his maximin, the outcome is satisfactory but is not an equilibrium.

These paradoxes do not hurt game theory if it is viewed as a purely formal theory. On the contrary, the conclusion that "rationality" becomes an ambivalent concept in certain contexts is a valuable discovery, having a bearing on the logic inherent in these situations. It is only the *prescriptive* aspect of game theory which is crippled by the paradoxes. A theory is not in a position to prescribe a rational decision if the meaning of rationality is not clear.

We suspect that these ambiguities stem from the peculiar forms of regress which characterize reasoning about someone else's reasoning, which, in turn, is based on assumptions about one's own reasoning, a point repeatedly stressed by Schelling (1960). In some types of games this process comes to an end in a finite number of steps, for example, in zero-sum games with a saddle point. In other types of games, even though the reasoning involves infinite regress, an extension of the strategy concept to include mixed strategies also leads to a satisfactory equilibrium. However there are still other games where this process leads to a paradox: if each player assumes that the other is individually rational, both can rationalize a strategy which is not collectively rational.

We suspect that the more fundamental source of the difficulties is an incompatibility between the fundamental assumption of game theory and the nature of "reflexive reasoning." The fundamental assumption of game theory is that everything there is to know about a situation is known at the start by "rational players," and "stays put" as they reason about the situation. Reflexive reasoning, on the other hand, "folds in on itself," as it were, and so is not a finite process. In particular when one makes an assumption in the process of reasoning about strategies, one "plugs in" this very assumption into the "data." In this way the possibilities may never be exhausted in a sequential examination. Under these circumstances it is not surprising that the purely deductive mode of reasoning becomes inadequate when the reasoners themselves are the objects of reasoning.

In the following chapter we shall admit the inductive mode of reasoning as an analytical tool of game theory. Whether the resulting investigations ought or ought not to be called game theory is, of course, a matter of terminological consensus. Some game theorists will probably feel that "inductive game theory" is properly not game theory, any more than experimental mathematics deserves the name of mathematics. The crucial factor to be introduced is that of an environment (including the other player) which is *not given* at the outset but which is to be *found out* in the process of playing the game, whereby one reacts to what one learns *and by these reactions modifies the environment.*

Admittedly this approach is more similar to that undertaken by mathematical psychologists in developing stochastic learning theories than to game theory. In fact, several investigators have already combined game theory with learning interactive concepts, including some of the game theorists themselves.[27] In my opinion this mixture leads to the most promising paths of development both for game theory and for behavioral science and therefore opens the way for developing a descriptive branch of game theory as an applied science.

10. *An Inductive Theory of Games: Dynamic Models*

It is commonplace to view the theory of games as the logical foundation of a theory of rational conflict. This view is justified in the context of games of pure strategy, of which Chess is a classical example. Indeed, the fascination which Chess has for the intellectual resides in the opportunity offered by this game for engaging in conflict conducted entirely on the symbolic level, where reason completely replaces physical prowess. Chess is a conflict which calls for complete ruthlessness and yet can be devoid of rage or hatred, emotions often felt by mature minds to be degrading to human beings.

From the game-theoretical point of view, Chess belongs to the class of games of pure strategy, not only in the sense that the degree of mastery of calculations and rational deduction completely determines the degree of skill in playing the game, but also in the technical sense of pure strategy. That is to say, Chess is a game with a saddle point, and this implies the existence of an optimal pure strategy available to each player. The definition of rational play in terms of the prescribed optimal strategy is the clearest and intuitively the most acceptable definition of strategic rationality.

We have seen how, in the case of games without saddle points, this straightforward definition of strategic rationality no longer suffices. It must be supplemented by the notion of maximizing expected gain or else by a supposition that a consistent rank order of preference can be assigned to all possible choices among risky outcomes (see p. 70).

In the context of nonzero-sum games, the definition of rationality must be further generalized. For example, if game theory is to prescribe a choice of strategy to each player in a nonnegotiable nonzero-sum game, the only *strategically rationalizable* optimal strategies are often not optimal in the sense of maximizing the payoffs of the players under the constraints of the game. Thus the definition of rationality bifurcates into individual rationality and collective rationality, and so unambivalent intuitively acceptable normative solutions are no longer available.

Once "rationality" acquires more than one meaning, we must specify context. In this chapter we shall define an inductive rationality, as contrasted with deductive rationality on which the entire formal theory of games has been based.

Classical logic distinguishes between deductive and inductive reasoning as follows: the former proceeds from general principles to particular instances, while the latter proceeds in the opposite direction. As is well known, mathematics admits only deductive reasoning, but natural science is impossible without some resort to induction. However, in the so-called exact natural sciences, persistent efforts are made to reduce the inductive steps to a minimum by replacing them with chains of deduction. The idealized goal of mathematical physics, for example, is to derive as many experimental facts as possible from as few general principles as possible. In physics, the role of mathematics is not confined to the quantitative description of data. Mathematics serves to forge theoretical physics into a compact deductive system.

However, as has been said, there is no natural science without an inductive component. Deductions are made from assumptions to conclusions. But the assumptions themselves must also be corroborated if the conclusions are to be accepted. The assumptions are assumed to be true either as generalizations of experience or on the basis of observations which corroborate but do not prove the conclusions. At any rate, generalizations based on observations must enter the process at some point, and so inductive logic must be invoked.

As long as game theory is developed as a purely deductive system of thought (like any branch of pure mathematics), it need not involve an inductive component. All the statements of game theory can be in the form "If so . . . then so" where the "if" part may remain forever hypothetical. However, as soon as game theory purports to be a prescriptive theory (purports to derive *optimal* course of action), the inductive component can no longer be dispensed with *in application*. For a given course of action is prescribed as optimal only on the basis of a given situation. The course of action recommended can be confidently undertaken if the situation assumed actually obtains. This can be established only by evidence of some sort, involving observations.

If induction is permitted as a component of game theory, the theory can be extended to situations where the game matrix is not known to the players. The players can nevertheless discover optimal strategies by trial and error, provided, of course, the results of their choices are made known to them, and they are given the opportunity to play a game many times.

Consider the following zero-sum game

	A_2	B_2
A_1	a	b
B_1	c	d

Game 37

where $a > d > b > c$. It is easily verified that this game has no saddle point (see p. 79). Therefore each player has an optimal mixed strategy:

$$x = \frac{d - c}{a + d - (b + c)} \text{ (the relative frequency of } A_1\text{);} \quad (78)$$

$$y = \frac{d - b}{a + d - (b + c)} \text{ (the relative frequency of } A_2\text{).} \quad (79)$$

If the players do not know the payoffs, they can only try to guess what such an optimal strategy might be. One way to guess it would be as follows. Let each try some arbitrary mixed strategy; say, player 1 tries x and player 2 tries y. When they use this pair of mixed strategies for some time, an average return will accrue to each, namely

$$G_1 = axy + bx(1 - y) + c(1 - x)y + d(1 - x)(1 - y) \quad (80)$$

to player 1, and

$$G_2 = -G_1 \quad (81)$$

to player 2.

Let now player 1 switch to some other mixed strategy, x' such that $x' > x$ (i.e., A_1 is used with greater frequency). The average payoff to player 1 resulting from x' may be either greater or less than the average payoff resulting from x. If it is greater, then player 1 knows that he is "on the right track." Consequently he will next choose x'', which increases the frequency of A_1 still more. If, on the other hand, the switch from x to x' brought in a smaller average payoff, then player 1 will conclude that he went "in the wrong direction" and consequently will choose an x'' which decreases the frequency of A_1 relative to x. Player 2 will be going through a similar trial and error search. Moreover, we may suppose that if a change in x brought in a large positive change in player 1's expected gain, the subsequent changes in x in the same direction will be large, and vice versa, and similarly for player 2. It is as if both players were driven toward their optimal strategies with a force which is stronger the greater the rate of change of expected payoff with respect to the rate of change of the frequency with which one or the other strategy is chosen.

This process can be described by a pair of differential equations,[28] namely

$$dx/dt = k_1 \frac{\partial G_1}{\partial x}, \tag{82}$$

$$dy/dt = k_2 \frac{\partial G_2}{\partial y} \tag{83}$$

where k_1 and k_2 are constants of proportionality. For simplicity, we shall take these constants to be unity. Then, taking the partial derivatives of G_1 and G_2 with respect of x and y respectively, we obtain the following pair of differential equations:

$$dx/dt = y(a + d - b - c) + b - d, \tag{84}$$

$$dy/dt = -x(a + d - b - c) + d - c. \tag{85}$$

Note that if we set $dx/dt = 0$ and solve for x and y, we get precisely the optimal mixed strategies x^* and y^* as the solution (see p. 81). However, this does not mean that the players using the trial and error method just described will necessarily "zero in" on the pair of equilibrium strategies. To see this, let us solve the pair of differential equations (84) and (85) to obtain x and y as functions of time. For simplicity of notation, denote $(a + d - b - c)$ by K.

The solution of the differential equations (84) (85) is

$$x = A \sin(Kt) + B \cos(Kt) + x^*, \qquad (86)$$

$$y = A \cos(Kt) + B \sin(Kt) + y^* \qquad (87)$$

where x^* and y^* are the equilibrium strategies given by (78) (79) while A and B depend on the strategies initially chosen. In the phase space[29] (x, y), the point which represents the (variable) pair of mixed strategies will be moving around a circle with (x^*, y^*) as its center. The magnitude of the radius will depend on the pair of strategies chosen initially. However, regardless of how large this circle will be, the *average* payoff accruing to the players will be G_1 and G_2, just as if they had chosen the equilibrium pair (x^*, y^*) to begin with.[30]

We see, therefore, that the players can, in principle, arrive at the "solution" of a mixed strategy game even if they do not know the game matrix to begin with but are only informed about "how they are doing" while they choose various mixed strategies by trial and error. In this case, then, an inductive approach leads us to the same result as a deductive one.

Let us now turn to a nonzero-sum game and see where a similar procedure leads us. Consider the following nonzero-sum game

	C_2	D_2
C_1	R, R	S, T
D_1	T, S	P, P

Game 38

where $T > R > P > S$.

Suppose we treat this game by the same method as the zero-sum game above (Game 37). Let C_1 and C_2 stand for mixed strategies in which the strategies so designated are chosen with frequencies C_1 and C_2 $(0 \leqslant C_1 \leqslant 1;$ $0 \leqslant C_2 \leqslant 1)$. Then $D_i = 1 - C_i$ $(i = 1, 2)$ and

$$G_1 = C_1C_2R + C_1(1 - C_2)S + (1 - C_1)C_2T + (1 - C_1)(1 - C_2)P. \qquad (88)$$

Taking the partial derivative with respect to C_1, we have

$$\frac{\partial G_1}{\partial C_1} = C_2(R - T) + (1 - C_2)(S - P). \qquad (89)$$

But according to our assumption $T > R$ and $P > S$. Therefore $\frac{\partial G_1}{\partial C_1}$ is always negative. Similarly it is easily seen that $\frac{\partial G_2}{\partial C_2}$ is always negative. Consequently, using trial and error method, the two players will use C_1 and C_2 less and less frequently until they use D_1 and D_2 exclusively.

But this is the equilibrium point of the game (considered nonnegotiable). Hence once more the inductive method leads to the same result as the deductive. We have obtained nothing new, and it seems so far that inductive reasoning simply corroborates the conclusions deduced from the knowledge of the game matrix. In a way this is gratifying, but in a way it is not. For recall that the conclusion regarding optimal strategies deduced from games like Game 38 was not satisfactory intuitively.[31] It led to a recommendation of a pair of strategies which was bad for both players. Is there a way of approaching the problem inductively that would lead to a different conclusion? Let us investigate further.

Consider two automata playing the following special case of Game 38:

	C_2	D_2
C_1	R, R	−T, T
C_2	T, −T	−R, −R

Game 39

where $R > 0$, $T > 0$.

Being very simple automata, they have no a priori knowledge nor any deductive ability. They simply react to what happens, and this in the simplest possible manner. Namely, if following a play of the game the payoff is positive, an automaton of this sort plays the same strategy on the next play. Otherwise he switches to the other strategy. Suppose also that at the initial play the strategies are chosen at random.

Then, if the initial choice happens to be (C_1C_2), it will remain so. If the initial choice happens to be (D_1D_2), both "players" will switch strategies because of the resulting negative payoffs. Consequently, the next outcome will be (C_1C_2) and will remain so thereafter.

Finally, if the initial choice is (C_1D_2), the first "player" will switch (because he has got a negative payoff) while the second player will repeat the same strategy (because he got a positive payoff). The next outcome, therefore, will be (D_1D_2). But this outcome will be followed by (C_1C_2), as we have seen, which will thereafter persist. The sequence following (C_2D_1) is analogous.

Thus we see that two automata which react only to whether payoffs are positive or negative will achieve "cooperation" in this game very easily. These automata, be it noted, are guided in their choices not by a simple propensity for the one or the other choice, but by a *conditional* propensity. For whether such an automaton will repeat, say, a C choice depends on what the other automaton did. If the other also played C, the first automaton will repeat it; but if the other played D, then the first automaton will not repeat the C choice. Similarly a D choice will be repeated, if and only if, the other chose C on that play.

We generalize the notion of conditional propensities by introducing the four variables, x, y, z, and w, defined as follows:

x_1 is player 1's propensity for choosing C
after a (C_1C_2) outcome;

y_1 is player 1's propensity for choosing C after a (C_1D_2) outcome;

z_1 is player 1's propensity for choosing C after a (D_1C_2) outcome;

w_1 is player 1's propensity for choosing C after a (D_1D_2) outcome.

Four other variables, x_2, y_2, z_2, and w_2, are defined analogously.

We now introduce an inductive model of this game. We assume that the players adjust not their unconditional propensities for the C or for the D choice (as in the case referring to Game 38), but rather adjust their conditional propensities x, y, z, and w in such a way as to maximize their expected payoffs.

Since the model will be presented here for illustrative purposes only, we shall simplify the situation drastically. Observe that in the case of the two automata discussed above, the values of the conditional propensities were implied to be the following: $x_i = w_i = 1$; $y_i = z_i = 0$ $(i = 1, 2)$. In the present model we leave $w_i = 1$; $y_i = z_i = 0$, but x_i is now a variable. In other words, we assume that each player will try to adjust his x (independently) so as to maximize expected payoffs.

We first calculate the expected payoffs to each player in terms of x_i and x_2. These turn out to be respectively[32]

$$G_1 = \frac{Rx_1x_2 + T(x_2 - x_1)}{2 + x_1 + x_2 - 3x_1x_2}; \tag{90}$$

$$G_2 = \frac{Rx_1x_2 + T(x_1 - x_2)}{2 + x_1 + x_2 - 3x_1x_2}. \tag{91}$$

Now differentiating G_1 and G_2 with respect to x_1 and x_2 and setting the derivatives equal to zero, we find the two strategies apparently in equilibrium:

$$x_1 = x_2 = x^* = \frac{T - R + \sqrt{R^2 + 7T^2}}{3T - R}. \tag{92}$$

Note: the solution involving the negative square root is discarded, because it implies $x^* < 0$, which has no meaning in the present context.

Does it then follow that the players, guided by trial and error adjustments of their propensities x_1 and x_2, will gradually bring these propensities toward the value x^* given by equation (92)? Not at all. This would be the case if the equilibrium in question were stable, that is, if small departures from the equilibrium set a process in motion which would bring the values x_1 and x_2 back toward the equilibrium position.

Let us examine the equation which represents both $\frac{\partial G_1}{\partial x_1}$ and $\frac{\partial G_2}{\partial x_2}$. In the two-dimensional space in which x_1 (or x_2) is plotted against $\frac{\partial G_2}{\partial x_2}$ $\left(\text{or } \frac{\partial G_1}{\partial x_1}\right)$ both equations represent the same parabola. The parabola is shown in Figure 8.

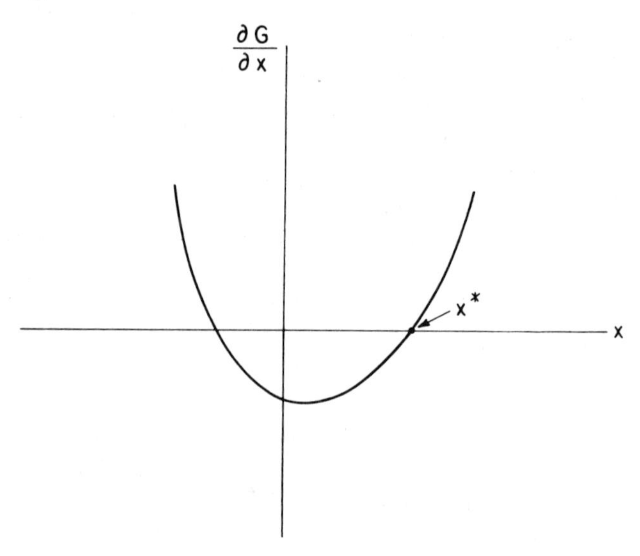

Fig. 8

Observe that the parabola intersects the horizontal axis twice. Of these two intersections only one lies between 0 and 1. Since x is a probability, this is the only intersection which has a meaning in the context of our model. Suppose now x_1 becomes slightly greater than x^*_1. The parabola rises above the horizontal axis, which means that $\frac{\partial G_2}{\partial x_2}$ becomes positive. But this means that x_2 will increase, because player 2, guided by trial and error, finds that he can increase his expected payoff by increasing x_2. But the rate of change of player 1's payoff with respect to x_2 is represented by the same parabola. Consequently, as x_2 increases, x_1 will also increase. *Mutual positive stimulation* sets in between x_1 and x_2, which drives the value of x_1 and x_2 toward unity.

Suppose, on the contrary, that x_1 has decreased to a value smaller than x^*. By the same argument, it can be shown that now mutual stimulation drives both x_1 and x_2 to zero.

Note: Even if $x_1 = x_2 = 0$, this does not mean that no cooperation will occur. It only means that following the outcome (C_1C_2) the outcome (D_1D_2) will inevitably occur. Because $w_1 = w_2 = 1$, (C_1C_2) will always follow (D_1D_2), so that if $x_1 = x_2 = 0$, the players will oscillate between (C_1C_2) and (D_1D_2), i.e., of the choices, fifty percent will be cooperative. However, this is a consequence of our special assumption that $w_i = 1$. A lower value of w_i will lead to lower values of C choices when $x_1 = x_2 = 0$. If $w_i = 0$, the players will "lock in" either on (C_1C_2) or on (D_1D_2), depending on the initial conditions.

We see, therefore, that this sort of trial and error adjustment model leads to an unstable equilibrium at $x_1 = x_2 = x^*$ if $x^* < 1$. Suppose this to be the case. If the values of x_1 and x_2 somehow become larger than x^*, they will be driven still further in the same direction toward unity. If they become smaller than x^*, they will be driven in the opposite direction, toward zero. The critical value of x^* is thus seen as a sort of threshold separating a trend toward maximum cooperation from a trend toward minimum cooperation. Consequently, this model of game behavior leads to results different from those prescribed by the purely deductive model, where everything is presumed to be known a priori, where the players are assumed to be rational and to strive to maximize their expected payoffs independently of each other.

The question before us is whether the trial and error adjustment procedure described above cannot also claim to be a rational procedure.

The procedure can certainly be called rational if complete knowledge of the game matrix by the players is *not* assumed. In the absence of such knowledge, the players must be guided by trial and error. Note that if they are so guided, then the long-run outcome depends on what parameters are being adjusted in the trial and error process. In particular, in the 2×2 zero-sum game without a saddle point, if the parameter being adjusted is the frequency of one of the strategies (in the case of just two strategies, this determines, of course, the frequencies of both strategies), the long-run result is identical to that prescribed by the game-theoretical minimax theory. In the nonzero-sum game examined, if the parameter being adjusted is the frequency of one of the strategies, again the long-term result coincides with the game-theoretical prescription of the equilibrium strategy. However, if the parameter adjusted is the *conditional* probability of response, the results are entirely different. They depend on the payoffs and on the initial conditions of the values of the adjusted parameters and exhibit a characteristic threshold effect. Roughly, this model, suggested for Game 39, leads to a conclusion that if initially the tendency of the players to "stick" with the (CC) outcome is sufficiently large, it will become still larger until (CC) outcomes occur exclusively. But if the initial tendency to stick with (CC) is not sufficiently large, it will become still smaller. Interpreted psychologically, this can be stated so: trust begets trust; distrust begets distrust.

Once we introduce psychological notions, we are outside the realm of game theory as it was originally formulated. The implications of this departure will be discussed more fully in the concluding chapters.

11. An Example: Inspector vs. Evader

We have now discussed the essential ideas which form the basis of the theory of the two-person game. We have used different games to illustrate the different principles: to define the concept of strategy, to reduce the game to normal form, to solve games with and without saddle points, to illustrate the way different bargaining principles are applied to a negotiated game, to exhibit paradoxes immanent in some nonnegotiated games, and to resolve these paradoxes by the use of an inductive theory.

We shall now discuss a single situation in which all of these principles can be illustrated. In thus relating the main problems of the theory to a model of a concrete situation, we hope to give the reader a "feel" for what is involved when one attempts to "apply" the theory, specifically an appreciation of both its conceptual power and its serious limitations.

Imagine an agreement concluded between two sovereign states, Urania and Plutonia, to the effect that each shall refrain from certain acts or activities. The treaty provides for an inspection procedure presumably designed to give assurance that the agreement is not violated.

Urania, for reasons which are irrelevant to our problem, insists on more inspection privileges, while Plutonia is reluctant to grant them. Obviously the entire problem, involving as it does a vast network of strategic, political, psychological, and possibly psychopathological components, is too complex to be formulated as a well-defined game. We can, however, isolate a salient feature of the problem and concentrate the entire conceptual apparatus of two-person game theory upon it. The result will be not a unique prescriptive solution of "the game"; for, as we shall see, even the drastically simplified version offers a large array of "solutions." The constructive feature of our result will be an unfolding of the problem into its various constituent parts and so a clarification of the issues, which ordinarily remain obscure.

Let us suppose that the issues have for a while revolved about the number of inspections to be allowed in a specified interval of time. As we have said, Urania has been pressing for more, while Plutonia has been holding out for fewer. At long last, however, the two sovereign states have agreed on the frequency of inspections. The only disagreement is on whether this frequency is to be realized by a *fixed* number of inspections per year or as an *average* number of inspections whose actual number can fluctuate statistically.

Specifically, suppose the year is divided into two six-month periods. Urania offers the following scheme. At the end of each period a coin is to be tossed. If it falls heads, an inspection is to be allowed; if it falls tails, not. In this way, although there may be one, two, or no inspections in any particular year, on the average there will be one inspection per year. Plutonia opposes this plan. She proposes a scheme whereby the question of whether there will or will not be an inspection in a six-month period shall be decided toward the end of that period by the inspecting party. If it is decided to inspect at the end of the first six months, then no inspection can take place during the last six months of the year. If it is decided to omit the inspection at the end of the first six months, then an inspection will be allowed during the last six months. (We shall also assume that the nature of prohibited activities is such that violations can be detected within six months but not after a longer period.)

From the positions of Urania and Plutonia, it appears that Urania is more interested in preventing violations by Plutonia than in having an opportunity for clandestine evasion. One might also suspect from Plutonia's position that she is less interested in inspecting Urania than in having an opportunity to evade. Let us therefore refer to the two high contracting parties as Inspector and Evader respectively.

The problem before us is which inspection schedule should Inspector prefer, the single allowed inspection per year or the two *potential* inspections per year, each with probability one half?

Urania's strategic experts argue that the probabilistic schedule is preferable to the fixed one. In support of their contention, they cite the so-called "end effect." For consider what is likely to happen if the fixed schedule is in effect. Clearly Inspector cannot always defer the inspection to the second period. For if he does, Evader can evade with impunity during the first. On the other hand, if Inspector randomizes his choice of period to inspect, Evader can simply wait until Inspector chooses to inspect the first period and then evade with impunity in the second. With a probabilistic inspection schedule, this is not possible, for whatever happened in the first period has no bearing on the second. Therefore Evader always stands the risk of being found out and is thereby deterred from evading.

The argument looks conclusive. Let us see, however,

how it looks in the light of game theory. Consider the following game consisting of four moves.[33]

Move 1. Evader chooses to evade or not in the first period.

Move 2. Inspector, not knowing how Evader has chosen (naturally, otherwise there is altogether no need for inspections!) chooses to inspect or not to inspect in the first period.

Move 3. Evader (knowing Inspector's choice on move 2 and, of course, his own choice on move 1) chooses to evade or not in the second period.

Move 4. Inspector, knowing the choices made in moves 1 and 2 (but not 3) chooses to inspect or not in the fourth period.

Inspector's knowledge of move 1 at move 4 is assumed to have come from some outside source, for example, espionage. It is also assumed that knowledge about evasion in a prior time period does not constitute a detected evasion. Thus if an evasion took place in the first period, which was not inspected, and an inspection took place in the second period when there was no evasion, then, although an evasion is credited to Evader, no detection is credited to Inspector.

Let Evader prefer the outcomes in this order: undetected evasion, no evasion, detected evasion. Next, we must assign the actual payoffs. For the time being we shall assume the game to be zero-sum, since it is often argued by the strategists of both Urania and Plutonia that the interests of Inspector and Evader, as well as those of the countries they represent, are diametrically opposed. Assuming, as usual, utilities on an interval scale, we assign $+1$ (to Evader) for successful evasion and 0 for no evasion. The payoff for detected evasion, therefore, must be negative. But we cannot specify this payoff without further information. We shall therefore leave it further unspecified and denote it simply by $-a$ ($a > 0$).

Finally, we shall assume that the payoffs are simply additive. Thus two undetected evasions shall be worth two utiles to Evader. Two detected evasions shall be worth 2a to Inspector, etc.

There are, as we have said, two versions of this game, namely under fixed inspection and under probabilistic inspection. We shall consider first the version under fixed inspection.

We examine the strategies available to each player. Inspector has only two feasible strategies: either to inspect in the first period or in the second. This is because under fixed inspection, he *cannot* inspect in the second period if he did so in the first, and he obviously *should* inspect in the second if he did not in the first.

Evader has, formally speaking, eight strategies of which, as we shall see in a moment, also only two are feasible. The eight formally possible ones are:

E_1. Evade in the first, and regardless of whether there has been inspection or not, also in the second.

E_2. Evade in the first and evade in the second if and only if there has been inspection in the first.

E_3. Evade in the first and evade in the second if and only if there has been no inspection in the first.

E_4. Evade in the first only.

E_5. Evade in the second only.

E_6. Refrain in the first; evade in the second if and only if there has been inspection in the first.

E_7. Refrain in the first; evade in the second if and only if there has been no inspection in the first.

E_8. Do not evade at all.

The game appears in normal form as Game 40, with Evader's strategies as above, and Inspector's strategies being I_1 (inspect in the first period); I_2 (inspect in the second period).

We see immediately that all of Evader's strategies except strategy 2 and strategy 6 can be eliminated from consideration. Strategy 2 dominates 1, 3, and 4; strategy

	I_1	I_2
E_1	$1 - a$	$1 - a$
E_2	$1 - a$	1
E_3	$-a$	$1 - a$
E_4	$-a$	1
E_5	1	$-a$
E_6	1	0
E_7	0	$-a$
E_8	0	0

Game 40

6 dominates 5, 7, and 8. Neither of the two strategies dominates the other. Thus the game is reduced to a 2×2 game:

	I	J
E	$1 - a$	1
F	1	0

Game 41

E: Evade in first period; F: Do not evade in first period.
I: Inspect in first period; J: Do not inspect in first period.

Note that the strategies in Game 41 refer to Evader's and Inspector's choices with regard to the *first* period. What happens thereafter is completely determined by the outcome of this game. For example, if Evader has evaded and been caught, then clearly he evades again in the second period, because under the fixed inspection schedule there can be no more inspections. Consequently, the payoff to Evader in this outcome is $1 - a$, the sum of the payoff for getting caught ($-a$) and for having got away with one evasion ($+1$), etc. Note also that F also counsels evasion in the second period if there has been an inspection in the first.

Next, we note that Game 41 has no saddle point and therefore calls for mixed strategies. We solve it by the usual methods described in Chapter 6. The solution prescribes the following:

Evader should evade in the first period with probability $1/(1 + a)$.

Inspector should inspect in the first period also with probability $1/(1 + a)$.

The expected payoff to Evader (the value of the game) in utiles is $1/(1 + a)$ and the negative of this quantity to Inspector.

We pass to the version played under probabilistic inspections. Here Inspector has nothing to choose. Not he but Chance decides whether there may be an inspection or not. If yes, then obviously he should inspect. If no, he cannot inspect. This version, therefore, is not a genuine two-person game (see p. 21) but only a "game against nature" in which Evader is the only bona fide player. Inspector merely represents Nature, which can be in each of four possible states with probability $\frac{1}{4}$. The probabilistic version is shown as Game 42. Evader has eight strategies as before. They have been labeled the same way. Nature has four strategies (i.e., states) represented by YY, YN, NY, and NN. Here Y stands for yes and N for no, in answer to the question whether an inspection is allowed in the first (second) period.

Now if the payoffs are actually utiles, then Evader should choose the strategy which maximizes expected payoff. We see, therefore, that if $a > 1$, Evader should never evade (should choose strategy 8). If, however, $a < 1$, he should *always* evade (choose strategy 1). In

	YY	YN	NY	NN	*Expectation*
E_1	$-2a$	$1 - a$	$1 - a$	2	$1 - a$
E_2	$-2a$	$1 - a$	1	1	$3(1 - a)/4$
E_3	$-a$	$-a$	$1 - a$	2	$3(1 - a)/4$
E_4	$-a$	$-a$	1	1	$(1 - a)/2$
E_5	$-a$	1	$-a$	1	$(1 - a)/2$
E_6	$-a$	1	0	0	$(1 - a)/4$
E_7	0	0	$-a$	1	$(1 - a)/4$
E_8	0	0	0	0	0

Game 42

the first case, his expected payoff is zero. In the second case, his expected payoff is $1 - a$.

We return to our original question: Which inspection schedule should Inspector prefer?

If Inspector's aim is to minimize Evader's payoffs, clearly he should prefer the probabilistic schedule, for under this schedule Evader's payoff is either 0 (if $a > 1$) or $1 - a$ (if $a < 1$). Both of these payoffs are smaller than $1/(1 + a)$, which is Evader's expected payoff under the fixed schedule. So in this context, Urania's strategists are right. But there are other formulations of the Inspector-Evader game.

It is instructive to examine a variant, in which the game ends as soon as an evasion has been detected. One interpretation of this version is that a provision in the treaty between the two countries allows unlimited inspection following a detected violation. Another interpretation is that the treaty is voided whenever an evasion is detected. It is assumed that the payoffs remain the same.

This version is shown in Game 43.

	I	J
E	$-a$	1
F	1	0

Game 43

Here Evader's best strategy (mixed) is to evade in the first period with probability $1/(2 + a)$. This is also Evader's expected payoff. Under the probabilistic schedule, the payoff matrix is modified only in the rows corresponding to Evader's strategies 1 and 2. Namely, the payoffs in row 1 become $(-a, -a, 1 - a, 2)$, so that the expected payoff is $3(1 - a)/4$. The payoffs in row 2 become $(-a, a, 1, 1)$ with expected payoff $(1 - a)/2$. If $a > 1$, this situation is identical with the preceding one from the game-theoretical point of view. If $a < 1$, Evader's maximum payoff under the probabilistic schedule is now $3(1 - a)/4$.

We have already seen that without assuming the termination of the game after detection, Inspector ought to prefer the probabilistic schedule. In that case the arguments of the Uranian strategists are justified. If, however, termination is assumed, it may be to Inspector's advantage to insist on the fixed schedule. This happens when

$$1/(2 + a) < 3(1 - a)/4, \tag{93}$$

which, in turn, is the case if $a < .46$ approximately.

So far, we have assumed that Inspector's payoffs are equal to those of Evader with the opposite sign. This is not necessarily the case, if Inspector's aim is to minimize the frequency of evasion. In this case Inspector's best strategy in Game 41 is still to inspect in the first period $\frac{1}{1 + a}$ of the time (assuming no termination), since this keeps evasions to a minimum. But now Inspector clearly prefers the fixed schedule to the probabilistic one if a < 1, since in this case Evader will *always* evade under the probabilistic schedule and only a part of the time under the fixed schedule. To be sure, Evader will get less payoff under the probabilistic schedule, but he will evade more than under the fixed schedule.

In this way game-theoretical analysis brings out clearly the distinction between the situation in which Inspector is interested in *catching* evasions and the situation in which Inspector is interested in *preventing* evasions. In the first case, Inspector should always prefer the probabilistic schedule if the game does not terminate with detection. If the game terminates with detection, Inspector should prefer the probabilistic schedule only if a > .46. In the second case (if there is no termination and if prevention rather than detection is the aim), Inspector should prefer the probabilistic schedule only if a > 1.

The distinction between the cases has a counterpart in law enforcement. If a law-enforcing agency has a positive payoff from arrests (e.g., fines, bonuses, promotions), and if it acts to maximize these payoffs, the result may not be a maximal reduction in the crime rate (assuming a zero-sum cops-and-robbers game). In some instances the ethical issues raised by this aspect of the gaming mentality are not serious. Consider the problem of fixing the level of fines for illegal parking. If the city's aim is to reduce illegal parking, it can do so by setting the fines sufficiently high. If, however, the city wishes to maximize the income from the fines, it will set the fines at a lower level, so as to encourage some violations. Such a practice does not seem very horrendous since no moral significance is attached to parking violations. It looks very different, however, when a police agent instigates a felony in order to add an arrest to his credit.

Coming back to our Inspector-Evader game, the view that the probabilistic schedule is *always* preferable from Inspector's point of view can be justified only if Inspector's aim is to maximize his own payoff (which in a zero-sum game includes rewards for catching evasions) rather than to minimize evasions.

How seriously this conclusion is to be taken is impossible to say in categorical terms. In my opinion, it should be taken at least more seriously than the so-called "solution" of the game. There is no question that the "solution" depends on ludicrously simplified assumptions, which may have deprived the model of even a modicum of reality. The distinction between the possible aims of Inspector, however, is not a "solution" and does not depend on the drastic simplification we have made. The distinction is a real life issue. Awareness of it does not depend, of course, on game-theoretical analysis. But we have seen how game-theoretical analysis *links* this issue to the distinction between the inspection schedules. In other words, if one is to justify a preference of one schedule over another on "hard" grounds, then one ought to derive such preference from "hard" analysis. If one derives an opposite result, then one must conclude either that wrong assumptions had been built into the game (which, of course, is quite likely), or else that one's preferences have a basis other than the conclusions of hard analysis (which is also a distinct possibility). This verdict comes out in the contradiction revealed by the analysis. For example, if $a < 1$, then one cannot maintain all three of the following assertions:

1. Inspector prefers the probabilistic schedule.
2. Inspector prefers to minimize evasions rather than to maximize one's payoffs in a zero-sum game.
3. The game as formulated so far is an adequate model.

At least one of these three statements must be false. The realization that this must be so can be credited to game-theoretical analysis.

Statement (3) above is, of course, very likely to be false. Let us, therefore, seek other models of the Evader-Inspector game.

The Nonzero-sum Model

To begin with, if the treaty between the two countries is to benefit both, and if Inspector represents the interest of his country, then we ought to suppose that a detected violation is worse than the absence of evasion not only for Evader but also for Inspector. Next, we may suppose that an inspection which proves the alleged Evader innocent should be worth more to Evader than a noninspected absence of evasion. In short, it is reasonable to suppose that Evader prefers the outcomes in the following order: successful evasion is better than inspection which clears him, which is better than no-evasion-no-inspection, which is better than detected evasion. Inspector, on the other hand, prefers the outcomes as follows: no-evasion-no-inspection is better than inspection-clearing-Evader,[34] which is better than detected evasion, which is better than undetected evasion.

These preference orders are also compatible with our previous model under the fixed inspection schedule, provided we assume that evading in the second period is worth less to Evader than evading in the first. For in that case, Evader's best outcome is (E, J), since it implies that he has gotten away with the valuable evasion; his next best outcome is (F, I), because after refraining in the first period, he can still evade in the second; his third best outcome is (F, J), since in that case he has refrained from evading in the first period but cannot evade in the second because inspection is sure to come; his worst outcome is (E, I), as before, assuming that the disutility of getting caught is greater than the utility of evading in the second period. By similar reasoning, we can verify that Inspector's preferences are $(F, J) > (F, I) > (E, I) > (E, J)$.

We shall now normalize the payoffs. Calculations will be simplified if we assign +1 to each player's best outcome and 0 to each player's next to worst outcome. Of the two remaining payoffs, then, one will be positive between 0 and 1, and one (representing the worst outcome) will be negative without restriction on its magnitude.

As in the preceding case, the strategic analysis leads to a 2×2 game, shown as Game 44.

	I	J
E	$-a, 0$	$1, -d$
F	c, b	$0, 1$

Game 44

Game 44 is a nonzero-sum game and therefore admits a variety of "solutions." We shall examine several of them.

Shapley's Solution

Suppose Game 44 is negotiable. (We shall discuss the "realism" of this assumption below.) We shall now apply Shapley's method to obtain a negotiated solution. There are two cases, depending on whether the point (c, b) is inside or outside the triangle joining the points (−a, 0), (0, 1) and (1, −d). The difference between the two cases is seen in Figures 9 and 10.

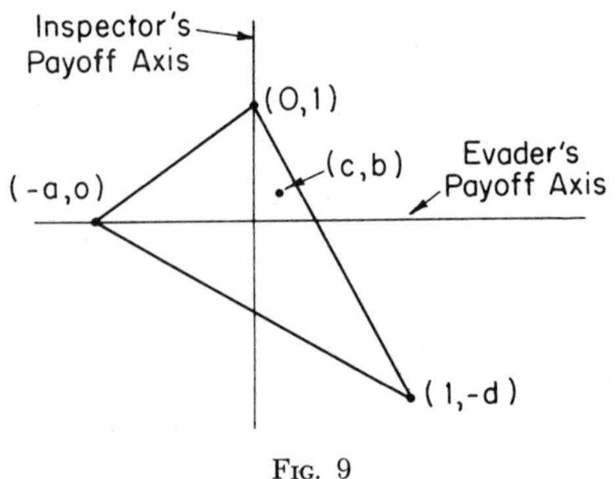

Fig. 9

If (c, b) is inside the triangle, the payoff polygon is the triangle and then the negotiation set is contained in

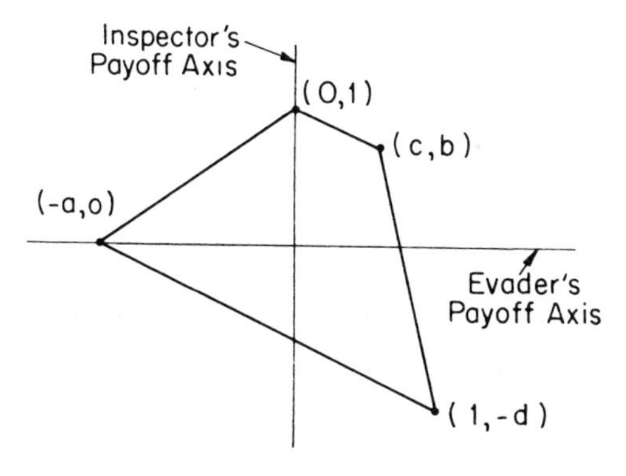

FIG. 10

the line joining (0, 1) to (1, −d). In the contrary case, the payoff polygon is a quadrangle, and the negotiation set is contained in the broken line joining (0, 1) to (c, b) to (1, −d).

Let us derive the conditions on the parameters which determine the one condition or the other. The equation of the line joining (0, 1) and (1, −d) is

$$Y - 1 = -(d + 1)X. \tag{94}$$

Setting $Y = b$ and solving for X, we obtain

$$X = \frac{1 - b}{d + 1}. \tag{95}$$

Here X and Y are the utilities accruing to Evader and Inspector, respectively. Now if $X \geqslant c$, when $Y = b$, the point (c, b) will be inside the triangle or on the boundary. If $X < c$, the point (c, b) will fall outside. Therefore, if the negotiation set is to be contained in the line $Y - 1 = -(d + 1)X$, we must have

$$c \leqslant \frac{1 - b}{d + 1} \quad \text{or} \quad b \leqslant 1 - c(1 + d). \tag{96}$$

We shall consider this case first.

From the game matrix, we find that Evader's security

level is $c/(1 + a + c)$,[35] which he can obtain by evading with probability $c/(1 + a + c)$ in the first period. As for Inspector, his payoff matrix has a saddle point at (E, I). Therefore, his security level is 0, which he can obtain by always inspecting (in the first period). Note that Evader's security level strategy is not best against Inspector's security level strategy. Nor is the latter necessarily best against the former. This is in consequence of the fact that the game is nonzero-sum. The reason for determining the pair of security level strategies is that they determine the status quo point in Shapley's negotiated settlement (see p. 108).

Accordingly,[36] we are able to maximize the quantity

$$\left[X - \frac{c}{1 + a + c}\right][1 - (d + 1)X] \qquad (97)$$

which gives us the negotiated solution

$$X^* = 1/2\left[\frac{1}{(d + 1)} + \frac{c}{(1 + a + c)}\right]; \qquad (98)$$

$$Y^* = 1/2\left[1 - \frac{c(d + 1)}{a + c + 1}\right]. \qquad (99)$$

If the solution is to be acceptable, X^* and Y^* must both exceed the respective players' security levels. But this can be easily seen to be the case because of inequality (96).

The above negotiated solution can be obtained if the players alternate between the outcomes (E, J) and (F, J). That is to say, Inspector agrees never to inspect in the first period. This means that Evader will never evade in the second period. He is allowed, however, to evade in the first period with probability

$$1/2\left[\frac{1}{d + 1} + \frac{c}{1 + a + c}\right]. \qquad (100)$$

Next consider the case when (c, b) falls outside the triangle, and consequently the payoff polygon is a quadrangle, as shown in Figure 10. In that case

$$b > 1 - c(d + 1). \tag{101}$$

Now the solution may be either on the upper or the lower segment of the broken line joining the points (0, 1), (c, b), and (1, −d). Suppose, first, that it lies on the upper segment. The equation of the line with which the segment coincides is

$$Y = 1 - \frac{1 - b}{c}X. \tag{102}$$

Accordingly, we must maximize

$$\left[X - \frac{c}{1 + a + c}\right]\left[1 - \frac{1 - b}{c}X\right]. \tag{103}$$

We obtain

$$X^* = 1/2\left[\frac{c}{1 - b} + \frac{c}{1 + a + c}\right]; \tag{104}$$

$$Y^* = 1/2\left[1 - \frac{1 - b}{1 + a + c}\right]. \tag{105}$$

If the solution is to be on the segment being examined, we must have $0 \leqslant X^* \leqslant c$. By inspection, we can see that $X^* > 0$. To have $X^* \leqslant c$, we must have

$$\frac{1}{1 - b} + \frac{1}{1 + a + c} \leqslant 2, \tag{106}$$

which after rearrangements reduces to

$$b \leqslant \frac{a + c}{1 + 2a + 2c}. \tag{107}$$

Combining inequality (107) with (101), we obtain the following inequality

$$\frac{a + c}{1 + 2a + 2c} \geqslant b > 1 - c(d + 1), \tag{108}$$

which is the condition sought. When this condition is fulfilled, the mixture is between the outcomes (F, I) and (F, J). That is to say, Evader never evades in the first period, while Inspector is obliged to inspect in the first period (therefore not the second) a certain fraction of the time. This fraction is computed so as to yield the payoffs given by (104) and (105). In other words we must find a θ such that $c\,\theta = X^*$. But this is

$$\theta = 1/2\left[\frac{1}{1-b} + \frac{1}{1+a+c}\right] \tag{109}$$

and so is the fraction of times that the first period is inspected.

Let now the solution lie on the lower segment, i.e., on the line joining (c, b) with (1, −d).

The equation of this line is

$$Y + d = -\frac{d+b}{1-c}(X-1). \tag{110}$$

We maximize

$$\left[X - \frac{c}{1+a+c}\right]\left[-d - \frac{d+b}{1-c}X + \frac{d+b}{1-c}\right]. \tag{111}$$

Using the same procedure as above, we obtain for this case

$$X^* = 1/2\left[\frac{-d(1-c)}{d+b} + 1 + \frac{c}{1+a+c}\right]; \tag{112}$$

$$Y^* = \frac{1}{2(1-c)}\left[\frac{acd + c^2d + ab + b}{1+a+c}\right]. \tag{113}$$

To satisfy the security level condition, we must have $0 < Y^* < b$. By inspection, $Y^* > 0$. To guarantee $Y^* < b$, we must have, as the reader can verify,

$$b > \frac{cd(a+c)}{a+1-2ac-2c^2} \quad \text{and} \quad a+1 > 2ac+2c^2. \tag{114}$$

Hence

$$b > \text{Max}\left[\frac{cd(a+c)}{a+1-2ac-2c^2}, 1 - c(d+1)\right]$$
$$\text{and} \quad a+1 > 2ac+2c^2. \tag{115}$$

When this condition obtains, the mixture is between (F,

I) and (E, J). Namely, Inspector will inspect the first period only when there is no evasion: Evader will evade in the first period only when there is no inspection.[37]

Nash's Solution

We shall confine ourselves to the case where the payoff polygon is a triangle (Figure 9), hence $b \leqslant 1 - c(d + 1)$, and so $Y = 1 - (d + 1)X$ is the equation of the negotiation set line. To determine the threat strategies,[38] we calculate the expected payoffs x_0 and y_0, resulting from the hypothetical pair of threat strategies x and y:

$$x_0 = -axy + x(1 - y) + c(1 - x)y, \tag{116}$$

$$y_0 = -dx(1 - y) + b(1 - x)y + (1 - x)(1 - y). \tag{117}$$

Once x_0 and y_0 are determined, the negotiated payoff to Evader is found by maximizing the quantity

$$[X - x_0][1 - (d + 1)X - y_0], \tag{118}$$

which determines

$$X^* = \frac{1}{2(d + 1)}[1 - y_0 + (d + 1)x_0]. \tag{119}$$

Evader wishes to maximize the quantity in the brackets of (119), while Inspector wishes to minimize it. Equivalently the quantity to be maximized (minimized) can be taken as $(d + 1)x_0 - y_0$. But these oppositely directed efforts of the players are equivalent to playing the following zero-sum game:

	I_1	I_2
E_1	$-a(d + 1)$	$1 + 2d$
E_2	$c(d + 1) - b$	-1

Game 45

The minimax strategies of Game 45 will be the threat strategies we are seeking.

What these minimax strategies are depends on whether Game 45 has a saddle point. We shall confine ourselves to that case. Game 45 has a saddle point, if and only if

$$b \geqslant (c + a)(d + 1). \tag{120}$$

To see this, observe that (E_1, I_2) cannot be a saddle point because $1 + 2d > -a(d + 1)$; (E_2, I_2) cannot be a saddle point, because $-1 < 1 + 2d$; (E_2, I_1) can be a saddle point only if $c(d + 1) - b < -1$, which contradicts the assumed condition $b \leqslant 1 - c(d + 1)$. Only $(E_1\ I_1)$ can possibly be a saddle point, provided

$$-a(d + 1) \geqslant c(d + 1) - b, \tag{121}$$

which is equivalent to (120).

Thus, if (120) holds, the threat strategies are pure strategies, namely E and I of Game 44. That is to say, Evader threatens to evade and Inspector threatens to inspect. The status quo point is $(-a, 0)$, and so (119) reduces to

$$\begin{aligned} X^* &= \frac{1}{2(d + 1)}[1 - a(d + a)] \\ &= 1/2\left[\frac{1}{d + 1} - a\right]. \end{aligned} \tag{122}$$

Then

$$Y^* = 1/2[1 + (d + 1)a]. \tag{123}$$

Note that $X^* > 0$, since otherwise $a(d + 1) > 1$, which violates (120). (Recall that $b < 1$.)

The Nonnegotiable Game

It may be argued that the Inspector-Evader game should be considered nonnegotiable since it is unrealistic to assume agreements between a would-be evader and an inspector concerning the number of “allowable” evasions. As we have said, whether this is a reasonable assumption or not depends on the social context with which game theory is not concerned. It is certainly in order, however, to investigate the nonnegotiable version of the game as well as the negotiable version.

First let us find the equilibria. We note that none of the four outcomes is an equilibrium, since one or the other player can always improve his position by switching to the other strategy, whatever the outcome. Let us see whether there is a mixed strategy equilibrium. Assuming strategy mixtures $(x, 1-x)$ and $(y, 1-y)$, we obtain the respective expected payoffs to Evader and Inspector:

$$\begin{aligned} G_E &= -axy + x(1-y) + c(1-x)y, \\ &= x(1 - ay - y - cy) + cy, \end{aligned} \tag{124}$$

$$\begin{aligned} G_I &= -dx(1-y) + b(1-x)y + (1-x)(1-y), \\ &= y(-1 + b + dx - bx + x) + 1 - (d+1)x. \end{aligned} \tag{125}$$

An equilibrium obtains if

$$x = \frac{1-b}{1-b+d}; \tag{126}$$

$$y = \frac{1}{1+a+c}. \tag{127}$$

Let us investigate the stability of the equilibrium. Suppose x deviates from (126) to a larger value. Then the coefficient of y in (125) becomes positive and Inspector can increase his payoff by increasing y. But if y increases to a value larger than the equilibrium value, the coefficient of x in (126) becomes negative, and consequently Evader can increase his payoff by *decreasing* x. In short, a positive deviation of x from the equilibrium leads to a restoring tendency back toward the equilibrium. The same can be shown with regard to the oppositive deviation of x and the deviations of y. Thus the equilibrium given by (126) and (127) is stable. The expected payoffs to Evader and Inspector are

$$G_E = \frac{c}{1 + a + c}, \tag{128}$$

$$G_I = \frac{db}{1 + b + d}. \tag{129}$$

We note that the equilibrium gives Evader no more than his security level, but it gives Inspector more than his security level.

A Dynamic Model

A full-scale dynamic model would involve all possible transition probabilities among the four outcomes. A drastically simplified model will suffice to illustrate the process. First, we introduce some notation. We shall designate by

α : the outcome (E, I),
β : the outcome (E, J),
γ : the outcome (F, I),
δ : the outcome (F, J).

We shall suppose the simplest type of dynamics. Namely, if in some outcome a player receives the payoff 1 (the highest), he will not switch his strategy on the next play. If in an outcome he receives a negative payoff, he will certainly switch. Otherwise he will switch with probability x (for Inspector) and with probability y (for Evader) but only if by switching he can improve his payoff.

These "rules" determine all the transition probabilities. For example, we can verify that the transition from α, i.e., (E, I) to γ i.e., (F, I) takes place with probability one. This is because Evader will certainly switch, since in α he gets a negative payoff. As for Inspector, he will not switch, because if he should (while Evader does not) his payoff will be diminished. Next, we can verify that the players will remain in γ with probability $1 - y$, or they will switch from γ to δ with probability y. This is because Evader cannot improve his payoff by switching from γ, but Inspector can. Similarly, we can verify the remaining transition probabilities, as shown in the following matrix:

	α	β	γ	δ
α	0	0	1	0
β	1	0	0	0
γ	0	0	$1 - y$	y
δ	0	x	0	$1 - x$

This stochastic process eventually leads to a steady state, in which the probabilities of the four states are given by

$$\alpha = \frac{xy}{x + y + 2xy}, \tag{130}$$

$$\beta = \frac{xy}{x + y + 2xy}, \tag{131}$$

$$\gamma = \frac{x}{x + y + 2xy}, \tag{132}$$

$$\delta = \frac{y}{x + y + 2xy}. \tag{133}$$

Accordingly, the expected long term payoff averages will be to Evader and Inspector respectively:

$$G_E = \frac{-axy + xy + cx}{x + y + 2xy}, \tag{134}$$

$$G_I = \frac{-dxy + bx + y}{2xy + x + y}. \tag{135}$$

As the players adjust the variables which they respectively control, G_E and G_I move about in (x, y) space. Whenever $\partial G_E/\partial x$ is positive, Evader adjusts x upward, and vice versa. Whenever $\partial G_I/\partial y$ is positive, Inspector adjusts y upward, and vice versa.

If there are equilibria in (x, y) space, they are obtained by setting these partial derivatives equal to zero. These derivatives turn out to be

$$\frac{\partial G_E}{\partial x} = \frac{y[c - y(a - 1)]}{(x + y + 2xy)^2}, \tag{136}$$

$$\frac{\partial G_I}{\partial y} = \frac{x[(1 - b) - x(d + 2b)]}{(x + y + 2xy)^2}. \tag{137}$$

There are four cases:

1. $a - 1 < c;\ d + 2b < 1 - b.$ (138)
2. $a - 1 > c;\ d + 2b < 1 - b.$ (139)
3. $a - 1 < c;\ d + 2b > 1 - b.$ (140)
4. $a - 1 > c;\ d + 2b > 1 - b.$ (141)

Case 1. Here both partial derivatives [(136), (137)] are always positive. Consequently both x and y increase until they reach their maximum values, i.e., 1. In this case a shift of strategy always occurs whenever a player can improve his position. The "system" keeps going through the four states (outcomes) counterclockwise. The average payoffs are

$$G_E = \frac{1 - a + c}{4};\ G_I = \frac{1 + b - d}{4}. \tag{142}$$

Case 2. Here $\partial G_I/\partial y$ is always positive. Consequently y keeps increasing and finally reaches 1. When this happens, $\partial G_E/\partial x$ is negative. Consequently, x keeps decreasing to zero. There results a stable equilibrium at $x = 0$, $y = 1$. Referring to [(130)-(133)] we see that the system comes to rest in state δ, i.e., (F, J). In this case Evader will not evade and Inspector will not inspect. The payoffs will be:

$$G_E = 0;\ G_I = 1. \tag{143}$$

Case 3. Here $\partial G_E/\partial x$ is always positive. Consequently x keeps increasing to 1. Then $\partial G_I/\partial y$ will become negative and y will drop to zero. The system will come to rest at γ, i.e., (F, I). In this case, Evader will never evade and Inspector will always inspect. The payoffs will be:

$$G_E = c; G_I = 1. \tag{144}$$

Case 4. Here an unstable equilibrium will obtain at $x = (1 - b)/(d + 2b)$; $y = c/(a - 1)$. Slight deviations of either x or y from this equilibrium will start x and y into opposite directions. Consequently the system will move either toward δ or toward γ. This case, therefore, reduces either to Case 2 or Case 3, depending on the initial conditions.

Probabilistic Inspection

To complete our investigation, we must investigate also the outcomes of the probabilistic inspection schedule. However, as we have formulated the game, we do not have enough information to treat this case, since under probabilistic inspection it is possible, for example, for Evader to evade twice and be caught both times, or not to be caught at all. Since this was not possible under the fixed inspection schedule, we have not assigned payoffs to some of the outcomes possible under the probabilistic schedule. Let us now see what we can infer and what we must assume in addition.

Let the following unknowns represent Evader's payoffs associated with the corresponding *single* events.

x: undetected evasion in the first period,
y: undetected evasion in the second period,
z: detected evasion in the first period,
w: detected evasion in the second period,
u: no evasion in the first period,
v: no evasion in the second period.

Now we can write

$$z + y = -a. \tag{145}$$

This is because $-a$ is Evader's payoff associated with the outcome (E, I), i.e., detected evasion in the first period which, under the conditions of fixed inspections, *implies* that Evader may evade with impunity in the second.

Similarly, examining the implications of the remaining outcomes, we can write

$$x + v = 1, \tag{146}$$

$$u + y = c, \tag{147}$$

$$u + v = 0. \tag{148}$$

The four equations [(145)-(148)] allow us to determine four of the single-event payoffs in terms of those associated with fixed inspections. Two payoffs, however, must remain undetermined. Accordingly, we must introduce two additional payoffs (Evader's):

e: the utility of undetected evasion in the first period;
$-f$: the utility of detected evasion in the second period,

where $e > 0$, $f > 0$.

Assuming the payoffs to be additive, we can determine the remaining payoffs, namely

$c - e + 1$: undetected evasion in the second period,
$-a - c + e - 1$: detected evasion in the first period,
$1 - e$: no evasion in the first period,
$e - 1$: no evasion in the second period.

Analogous calculations of Inspector's payoffs yield the following:

$-g$: undetected evasion in the first period,
$b - d + g - 1$: undetected evasion in the second period,
$-b + d - g + 1$: detected evasion in the first period,
h : detected evasion in the second period,
$d - g + 1$: no evasion in the first period,
$-d + g$: no evasion in the second period.

Here we have introduced $-g$ and h ($g > 0$, $h > 0$) as additional payoffs. From Evader's utilities, we conclude the following.

Evader will evade in the first period if

$$\frac{-a - c + 2e - 1}{2} > 1 - e \tag{149}$$

or if

$$4e - a - c - 3 > 0. \tag{150}$$

Evader will evade in the second period if

$$\frac{-f + c - e + 1}{2} > e - 1 \tag{151}$$

or if

$$3e - 3 + f - c < 0. \tag{152}$$

Since under probabilistic inspections Inspector no longer makes decisions, Evader's choices in the first and second period are independent. That is to say, he will evade in both the first and second period if inequalities (150) and (152) are both satisfied, and will never evade if neither of those inequalities is satisfied. A necessary condition that there be no evasions at all is

$$4e - a - c - 3 < 3e - 3 + f - c \tag{153}$$

or

$$e < a + f. \tag{154}$$

If there are no evasions, Inspector gets the largest possible payoff, 1. He will therefore prefer the probabilistic schedule to any other arrangement, including all the forms of negotiation.

Note, however, that (154), while necessary, is not sufficient. If either inequality (150) or (152) is satisfied, it is worth Evader's while to evade either in the first period or in the second. In that case, depending on the payoffs to Inspector, the latter may prefer either the fixed schedule or the probabilistic one.

As an example, suppose Evader's payoffs are such that

he always evades in the first period. Then Inspector's expected payoff turns out to be $(1 - b - d)/2$. If we compare it to, say, what Inspector expects at the equilibrium of the nonnegotiated game with fixed schedule, namely $bd/(1 + b + d)$ [see (129)], we see that it may very well happen that the latter payoff is larger.

In conclusion, it appears that Inspector's preference for the one or the other schedule is by no means obvious. If the problem is analyzed quantitatively, taking actual utilities into account, we see that within certain ranges of utilities the fixed schedule may well be to the Inspector's advantage. The commonsense argument in favor of the probabilistic schedule really contains a tacit assumption either to the effect that the game is zero-sum or to the effect that the disutility of being caught in an evasion is sufficiently large to prevent evasion altogether under the probabilistic schedule.

We now are in a position to say under what conditions Inspector will prefer fixed inspection schedules (i.e., the Uranian strategists are mistaken in their conclusion that probabilistic inspection is always preferable).

We shall confine ourselves to comparing Inspector's payoffs under probabilistic inspections with those under fixed inspections when the game is not negotiable, specifically with the equilibrium solution to that game.

1. If $c + 3 < \text{Min}\,[4e - a, 3e + f]$, i.e., if Evader always evades in the first period, never in the second, then fixed inspections are preferred by Inspector if

$$\frac{bd}{1 + b + d} > \frac{1 - b - d}{2} \tag{155}$$

or

$$4bd > 1 - b^2 - d^2. \tag{156}$$

2. If $c + 3 > \text{Max}\,[4e - a, 3e + f]$, i.e., if under probabilistic inspection Evader always evades in the first period, never in the second, then fixed inspections are preferred by Inspector if

$$\frac{bd}{1+b+d} > \frac{1+b+d+h-g}{2} \qquad (157)$$

or

$$2bd > (1+b+d)^2 + (h-g)(1+b+d). \qquad (158)$$

3. If $3e + f < c + 3 < 4e - a$, i.e., if under probabilistic inspection Evader always evades, then fixed inspections are preferred by Inspector if

$$\frac{bd}{1+b+d} > \frac{2h+d-4g-b}{4}. \qquad (159)$$

Obviously this condition will be more easily satisfied than (157), since the right side of (157) is larger than that of (159) for all values of the variables.

4. If $4e - a < c + 3 < 3e + f$, i.e., if under probabilistic inspections no evasions will ever occur, then probabilistic inspection is, of course, preferred.

These results are merely quantitative statements of the commonsense conclusion that probabilistic inspections will not stop evasions if it is worthwhile for Evader to evade even if he gets caught now and then. If so, then under probabilistic inspections he will always evade. The fixed inspection schedule, on the other hand, induces a genuine two-person game, in which Evader can maximize his payoff by judiciously randomizing evasions, i.e., by evading only part of the time. If the game is nonzero-sum, Evader's optimizing strategy may actually be of benefit to Inspector as well.

12. Opportunities and Limitations

Of what use is game theory? Questions of this sort are always put to the theoretician. The theoretician sometimes resents such questions because he senses behind them a suspicious attitude toward any activity not directed toward one of very few goals, such as accumulation of wealth or of power, or of technology, the latter being frequently regarded as subservient to the former. The theoretician often gives vent to his resentment by championing the cause of "pure science." Sometimes he goes so far as to declare that application degrades science. Mathematicians especially, being dependent in their creative work on almost absolute freedom of choosing a postulational framework (for this is the way original mathematical systems are created), sometimes take pride in the fact that their mathematical theories seem extremely remote from application.

In order to examine these matters without prejudice, the emotional connotations of the pure vs. applied science controversy should be suppressed at least until the meanings of the terms have been clarified. It is reasonable to respond to the question "Is this theory useful?"

with a counter question "Useful for what?" If it indeed turns out that the "usefulness" refers to crude ambitions and appetites, like moneymaking or acquisition of power, or reveals a parochial view of science as a handbook of gadgetry, then perhaps a simple "No" to the first question is in order, even if this answer is not entirely correct. A negative answer in this case could be given as a paraphrase of a comment which one might not wish to make directly: "Maybe the theory is 'useful,' but if those are the uses you are thinking of, I couldn't care less."

There is no reason, however, why the theoretician should share the shortsighted view of what is "useful" or "applicable." For example, many branches of abstruse mathematics are eminently useful because they shed light on other (perhaps equally abstruse) branches of mathematics. In this way the development of mathematical theory serves the purpose of unifying the various branches of mathematics by revealing the logical connections or analogies among them. One may, of course, ask whether the unification of mathematics is useful, and if so, how. To this one might reply that any theoretical unification serves the purpose of "compressing" knowledge so that more of it can be fitted into a single mind, and this, in turn, facilitates the further expansion of knowledge. One then invites the question of whether the expansion of knowledge is useful, to which one could answer unequivocally "Yes," which is simply a value judgment. Arguments about value judgments are notoriously sterile. Still, this sort of inquiry brings out distinctions between unsystematic accumulation of knowledge (which is perhaps of questionable value beyond a certain point) and a highly organized expansion of knowledge, i.e., enlightenment, which is a very different thing.

The question of the "usefulness" of game theory is extremely important because of serious misconceptions concerning the range of applicability of the theory. These misconceptions are in large measure due to the social climate in which game theory was developed.

As we have already pointed out, game theory can be viewed as an extension of a theory of rational decision to situations characterized by conflicts of interest. As such, the development of game theory *appears* to parallel the development of the theory of probability whose origin was in the analysis of games of chance. For example, the British philosopher R. B. Braithwaite (1955) wrote:

> No one will doubt the intensity, though he may dislike the color, of the (shall I say) sodium light cast by statistical mathematics, direct descendant of games of chance, upon the social sciences. Perhaps in another three hundred years' time economic and political and other branches of moral philosophy will bask in radiation from a source—theory of games of strategy—whose prototype was kindled around the poker tables of Princeton.

Now the theory of probability is today a mature branch of mathematics with a tremendous range of application in the physical, biological, and social sciences. However, it is also possible to view the theory of probability as an extension of a theory of rational decision to situations in which the outcomes of choices of action are uncertain. This is the context in which the theory of probability was originally developed. In a game of chance one can, in the simplest case, suppose that one is faced with the choice of accepting or rejecting a bet on an event or, somewhat more generally, of choosing among alternative bets. The concept of expected gain solves with one stroke all such problems, provided either (1) the utilities of the payoffs are linear functions of the payoffs or (2) the bet in question is offered so many times that in the long run the expected average payoff becomes the actual average payoff.

Certain business decisions can also be viewed as bets. If the probabilities of alternative outcomes are known, and if the situation in question presents itself sufficiently many times, then the principle of rational decision which emerges out of probability theory can be applied. The applicability of the theory to decision problems is then established.

One can now argue by analogy that if two-person game theory is an extension of rational decision theory to situations in which outcomes are controlled by two decision-makers whose interests are at least partially in conflict, then the range of application of two-person game theory ought to be the range of such situations. It is understandable why, in the period following the close of World War II, when so much attention was paid, especially in the United States, to the impending power struggle between the Communist and the non-Communist worlds, the appearance of game theory on the scientific horizon was hailed with enthusiasm and with great expectations.

People had witnessed the increasing abstruseness of the sciences geared to military applications. World War I had been called the chemists' war. World War II was called the physicists' war. Toward its final phases, World War II was rapidly becoming a mathematicians' war with cybernetic devices and electronic computers beginning to play a decisive role. It is assumed in many quarters that World War III (which many feel to be a matter-of-fact culmination of existing trends) will be truly a mathematicians' war.

Moreover, mathematics is assumed in those quarters to be not merely an appendage to physical science but also the foundation of strategy. "Postures" are frequently calculated in terms of destructive potential, which, in turn, is calculated from projected "nuclear exchanges" and capabilities.

Reading the writings of nuclear age strategists, one often gets the impression that strategy as a science is enjoying a renaissance. Formal military strategy had a golden age during the European wars of the eighteenth century. These were primarily wars of maneuver; and military specialists took great pride in conducting their campaigns in accordance with classical strategic principles.

The first large twentieth-century war must have been an enormous disappointment to those specialists. With maneuver in open country made practically impossible by the formidable fire power of the machine gun and of artillery, World War I degenerated into a war of attrition without intellectual interest. World War II, however, was again largely a war of movement, and consequently interest in strategy and maneuver was revived.

Wars to come are imagined by the strategists to be either "limited wars" or "nuclear exchanges," both being envisaged as wars of strategy rather than of attrition. It seems that those strategists who are actively concerned with the conduct of limited war view such wars as "rational" instruments of national policy, in contrast to nuclear war which, because of its awesome destructiveness, falls outside the scope of rational policy. Those strategists, on the other hand, who are concerned with nuclear "exchanges," although noncommittal about the "rationality" of such maneuvers, view the *potential* for waging nuclear war as bargaining leverage in international affairs. They view the use of this potential as a basis of a rational diplo-military policy.

Both schools of strategy assume that the use of force, and certainly the threat of force, are instruments of rational conflict. Both place great emphasis on rationality in the conduct of conflict. In fact, Karl von Clausewitz, the author of the classical treatise on strategy, *Vom*

Kriege, appears often explicitly as a hero in the writings of contemporary strategists. For Clausewitz was a champion of the rational use of force. War should never become an activity for its own sake, Clausewitz insisted. Wars should be fought in the pursuit of specific goals, which were usually understood as opportunities to enhance the power of the war-waging nation vis-a-vis rival nations. Accordingly, Clausewitz viewed war as a civilized, intellectually challenging enterprise. He probably would not confer the dignified title of war upon the massacres among "savages," and it is an open question what he would have thought of the "total wars" of our century.

In summary, then, one finds in the writings of contemporary strategists a deliberate striving to rehabilitate war as a normal event among civilized nations. The reestablishment of high intellectual content in military strategy doubtless serves this purpose. In my opinion, the tremendous interest aroused by game theory is in no small measure due to the climate in which the rehabilitation of war, or at least of the sophisticated power struggle, was undertaken.

It becomes, therefore, extremely tempting to those actively involved in game theory and also interested in its application potential to reply in the affirmative to the question "Is game theory useful?" Since rationality in conflict enjoys extremely high prestige in our day when "realism" and "tough-mindedness" are extolled as evidence of sophistication and maturity, game theory can indeed be sold as a useful science.

To the credit of the game theoreticians (and by those I mean the genuine ones, those who understand the logical structure and the technicalities of game theory), they seldom, if ever, present game theory in this light. If pressed, they may cite a few illustrative examples; but they eschew claims to the effect that a formal training in game theory will soon become as necessary to a military strategist as a formal training in physics is necessary to an engineer. Sanguine prognoses for game theory as an adjunct to military science are usually traceable to the military profession and to its fringes, not to the game theoreticians. The military man with a faith in "science" would *like* to have a way of coming by strategic know-how "scientifically." From this sort of wishful thinking arose the great hopes centered on game theory and manifested in the support of research and the financing of conferences on the subject. The topic makes good newspaper copy, too.

In what follows we shall list what is required in order to develop game theory into an applied science in the field of application with which it is frequently associated. As we have said, the listing of analogous requirements for probability theory is a comparatively simple matter. If the field of application of probability theory is gambling, whether in casinos or in business, if the payoffs are in money, if utilities are linear with money, or if the bets are presented sufficiently frequently, and if the probabilities of all the possible outcomes are known, then probability theory provides a general method of solving all decision problems, which, in the context of gambling, are all of the same sort: choices among bets (including no bet). To make use of probability theory outside of money-betting gambles, it is also necessary to assign utilities on an interval scale to all the possible outcomes. Let us now see what the situation is in game theory.

First, let us take the simplest case, namely a well-defined two-person zero-sum game of strategy (e.g., a parlor game) played for money, the utility of money being a linear function of the amount. In this case the problem of assigning utilities does not arise. The game-theoretical problem appears in its purest form, namely that of finding optimal strategies (pure or mixed) for both players in the sense of maximizing expected payoffs under the given constraints.

As long as game theory is viewed as a "pure" science, the minimax theorem, which merely asserts the existence of optimal strategies, appears as an eminently important finding. A different demand, however, is put on applied science. Applied game theory must indicate actual solutions of actual games. The enormous difficulty of finding such solutions is evident from the fact that it is all but humanly impossible to obtain either the matrix representation or the game tree of any parlor game worth playing. In the case of a very simple game like Tic-Tac-Toe, a game tree could conceivably be drawn, but a matrix representation of the strategies remains out of the question. The number of strategies of Tic-Tac-Toe is enormous even if the symmetries of the game are taken into account (see p. 43).

On the other hand, the game of Tic-Tac-Toe probably was solved in the practical sense already in antiquity, without the benefit of game theory. The solution becomes clear in the course of repeated plays to anyone with a mental age of about eight. While the game-theoretical conclusion about the outcome of every play of Tic-Tac-Toe is of great value in giving us insight into the nature of a certain class of games, it is hardly an accomplishment of an applied theory. The value of an applied theory is estimated from its power of finding solutions *which had eluded the practitioners*. Game theory has few successes on this score. (Perhaps the solution of Morra [see p. 91] could be cited as an exception.)

The situation is different if a matrix representation of a decision problem can be obtained. As we have seen, this cannot generally be done for parlor games. It can, however, be done in situations where each of two decision-makers has a comparatively small number of decisions at his disposal. Here is where the full power of game-theoretical analysis is revealed. Especially in situations represented by games without saddle points, game theory has made a genuine and original contribution in the sense of a prescriptive theory. In general, such problems would have remained unsolved without the game-theoretical notion of mixed strategy and without the methods of computation, likewise discovered by game theoreticians, of arriving at the optimal mixed strategies.

Let us, therefore, see what must be known in order to make use of this important game-theoretical result in real situations. Situations in which each decision-maker has a choice of only a few strategies, although not found in the usual parlor games, are frequently found in real life. Businessmen, administrators, and military strategists frequently must decide between a few alternative courses of action or plans in situations in which the outcomes of decisions are determined not only by chance but also by the choices made by other decision-makers, whose interests in the outcomes are, in general, different. Here, then, is the sort of situation to which the fundamental game-theoretical result on two-person zero-sum games might be applicable, provided, of course, the "games" in question are indeed zero-sum.

If a game of this sort is indeed zero-sum, the problem of assigning utilities to the several possible outcomes reduces to the problem of assigning only one's own utilities. The utilities of the other are given as soon as one's own are determined, because in a zero-sum game the former are equal to the latter with the sign reversed.

How shall the decision-maker assign his own utilities to the outcomes? If the game has a saddle point (hence a pure strategy solution), the utilities need to be assigned on a scale no stronger than the ordinal scale. This sort of assignment is easiest. It requires only that the decision-maker rank all the outcomes in his order of preference of them without regard for the sizes of the intervals between the assigned preferences. Therefore it is reasonable to begin the analysis by rank ordering the outcomes. Then the outcomes are placed into the proper boxes of the game matrix (each being the result of a pair of strategy choices by the two players). If it turns out that some outcome is at the same time the worst in its row and the best in its column, a saddle point is determined. In that case, the player need not worry about ascribing more exact magnitudes to his preferences. The ordinal scale is sufficient. It only remains to choose a strategy containing the saddle point, and the problem is solved.

Yet, the typical real-life decision problem is usually much more involved. For the time being, we shall continue to assume that there are comparatively few strategies to choose from (not an unrealistic assumption) and that the strategies open to the other player are also known (a somewhat less realistic assumption). However, we have actually assumed more. We have assumed that the outcomes are determinate events. This is not generally the case in real life. Even if the strategies are determinate, i.e., pure, the outcomes may well be not determinate events but "lottery tickets." For example, let the choices open to a military commander be (1) to attack Sector A or (2) to attack Sector B; while the corresponding choices of the opposing commander are (1) to reinforce Sector A or (2) to reinforce Sector B. It is too much to expect that the result of a pair of strategy choices will be a perfectly determinate one, for example, a breakthrough with so many casualties, or a failure of breakthrough with so many casualties. Realistically, at most probabilities can be attached to success or failure and to each possible casualty figure. Let us assume that the expected number of casualties can be estimated from experience in similar situations. Then the outcome remains as the probability of a breakthrough at such and such (estimated) casualty cost. The game matrix (of outcomes, not utilities) might then look like this:

	A_2	B_2
A_1	.30 (4)	.60 (10)
B_1	.50 (3)	.40 (6)

Game 46

The chances of breakthrough and the estimated casualties (in thousands) in each of the four contingencies when the row chooser attacks the corresponding sector and the column chooser reinforces one or the other.

Now the attacking commander must undertake to rank the four outcomes in the order of his preference. Among the outcomes (A, A), (B, B), and (B, A) the first choice is easy. Certainly (B, A) is preferred to both (A, A) and to (B, B) because the probability of breakthrough is larger in (B, A) and the estimated casualty costs are smaller than in either (A, A) or (B, B). But how about the relative utilities of (A, A) and (B, B)? To compare these, estimated casualties must be weighed against the probability of breakthrough (also estimated). It makes a difference how preference is assigned. The problem also arises when (A, B) is compared with (B, A). The way the preference will be assigned will make an essential difference in the resulting decision problem. Suppose first that the commander considers ten thousand (estimated) casualties altogether prohibitive, so that he prefers (A, A) to (A, B). He also thinks that additional two thousand casualties are not worth an increase of .10 in the probability of a breakthrough, and accordingly prefers (A, A) to (B, B). Then the game matrix with utilities assigned on an ordinal scale will look like this:

	A_2	B_2
A_1	3	1
B_1	4	2

Game 47

This game has a saddle point, namely at (B, B), and the commander's choice is clear. He should attack Sector B. Suppose, on the contrary that the commander is relatively insensitive to casualty costs. He thinks that it pays to have two thousand excess casualties to increase the probability of breakthrough from .30 to .40, and that it is worth ten thousand casualties to have a better than 50-50 chance to break through (A, B). Then the game matrix will look like this:

	A_2	B_2
A_1	1	4
B_1	3	2

Game 48

This game has no saddle point. Its solution involves the use of mixed strategies. Now it is no longer sufficient to assign utilities on an ordinal scale. They must be assigned on an interval scale if the concept of optimal mixed strategy is to be meaningful. This means that the commander must specify how many casualties it is worth to him to increase the probability of breakthrough by any specified amount. This exchange ratio may be different in the different ranges of casualties or probabilities. For instance, it may be worth more casualties to increase the probability of a breakthrough from .50 to .60 than from .30 to .40. Or else the marginal disutility of a casualty may be different in the high casualty range from what it is in the low casualty range, etc.

In short, if the game theoretician is to solve a decision problem of the sort we have examined, he must have the sort of information we have described. Can decision-makers supply such information aside from making arbitrary on-the-spot estimates on the basis of hunches? We think not.

Now the concept of utility, as it appears in game theory, is developed in a manner which frees the decision-maker from assigning numerical utilities to the outcomes. All that is required from him is a rank ordering of the outcomes, provided, however, that risky outcomes are included in the ordering. Thus, in order to construct a game matrix with utilities on an ordinal scale, the game theoretician needs only to know how the decision-maker rank orders the associated risky outcomes. For example, given that the preference rank order among the four outcomes is (A, B) > (B, A) > (B, B) > (A, A), the decision-maker must answer questions of this sort: "Which do you prefer, (B, B) with certainty, or a 50-50 chance between (A, B) and (A, A)?" and so on, for all possible combinations of outcomes and associated probabilities. Recall that the outcomes themselves are *already* risky outcomes, i.e., expectations associated with events rather than events themselves. Thus, in constructing an interval scale, estimates must be made of "expectations of expectations." It is highly unlikely that questions of this kind can be understood by people involved with real life decisions, let alone answered with any degree of assurance. We thus see that even in the simplest real life situations a meaningful application of game theory is beset by formidable difficulties.

We have not nearly exhausted the sources of these difficulties. We have not yet raised the question of how probabilities of events are to be assigned in the first place.

The assignment of a probability to an event depends on a definition of a "universe" of events, i.e., a listing of all the different ways in which an event can happen, these ways being a priori equally likely. The event whose probability is to be determined is viewed as a certain combination of these basic events. For instance, a roll of a die can result in 1, 2, 3, 4, 5, and 6. The event "Even, not greater than 5" comprises the elementary events 2 and 4, and accordingly has probability $\frac{1}{3}$.

These results are simple consequences of the definition of probability. To apply the concept of probability to real life decisions, we must have some empirical justification for the way probability is defined. Such justification is actually obtained if identical situations recur many times. For instance, if we roll a fair die many thousands of times, the outcome "Even, not greater than 5" will in fact occur with a frequency which is very nearly one third of the times.

Some events are by their very nature nonrepeatable. For instance, a boxing match takes place between two individuals who, as a rule, meet only once or at most very few times. It is therefore impossible to estimate the probability of victory for the one or the other on the basis of observed frequencies of such victories. How, then, are such probabilities estimated? We know that they *are* estimated, because the estimated probabilities are reflected in the odds offered or accepted on bets about the outcome of the match.

One is tempted to guess that such estimates are no more than personal hunches. Nevertheless a certain consensus often exists among people professionally concerned with bets of this sort. Therefore, "factors" must somehow be taken into consideration and weighed. It is very difficult to make these considerations explicit. Professional gamblers may talk knowingly of what they take into account when they bet on sporting events, but these explanations can hardly be translated into numerical computations. They are seldom more than elaborations of their feelings or beliefs.

When events are repeated, there is more opportunity for objective estimates. For example, in the course of a baseball season, each team will play every other team of its league many times. In the course of several seasons the number of games between two specific teams may form a statistically meaningful sample. Or rather, the sample would be statistically meaningful if the conditions under which the games were played were identical. We know, of course, that this is not the case. The composition of the teams changes. Whether a team plays on its home grounds or not is said to make a difference. The weather, the time of the season, and the team's past experience both immediate and cumulative also probably make a difference. In trying to estimate the probability of victory of one team over the other, we are faced with a dilemma: on the one hand, it is desirable to have a large sample of games between the teams for a reliable estimate; on the other hand, it is desirable to have the conditions under which the games are played as constant as possible. If we try to achieve the latter, we sacrifice the former, because if we try to standardize the conditions, we reduce the number of events in each category.

In short, the estimate of probabilities of real life events is not at all a simple matter. When the nature of probability is carefully examined, it appears as if the notion of probability as some sort of objective attribute of an event evaporates. A probability, we have seen, depends on the universe of events in which the event in question is imbedded. But it is largely up to us to choose that universe. Therefore a subjective element is always present in the estimate of probabilities, and this subjective factor cannot be eliminated unless the probabilities are checked against observed frequencies; and this cannot be done if the probability to be estimated relates to an essentially unique event.

Coming back to our commander, we had credited him with the ability to estimate expected casualties. Presumably he does this by reference to experience, the relevant experiences being those related to similar situations. But the degree of similarity of situations is always open to question, and the experience of a single person is only a small sample of possibly related events. Therefore we were extremely generous in taking the commander's estimate seriously in an objective sense. At most his estimates reflect his own degree of belief. In all likelihood, if he is forced to assign numbers both to probabilities and to utilities, he will do so either arbitrarily or in such a way that his intuitively preferred decision is rationalized by the assignment. Under these circumstances, it is difficult to view game theory as a tool for rendering decisions precise and rational. Nor is there any evidence that decisions based on game-theoretical calculations lead to measurably better results than decisions based on commonsense considerations intuitively arrived at.

Nothing of what has just been said applies to situations where probabilities and utilities can be determined with reasonable precision, and where the outcomes are determined by a single decision-maker. The field of operations research, for example, is largely concerned with well-defined optimization problems, where optimization depends on finding a maximum or a minimum of a well-defined quantity, be it efficiency (precisely defined) or cost or some combination of these, or even a probability (provided the situation presents itself many times as, for example, in problems of quality control). Problems of this sort can be meaningfully solved in terms of the values assigned. Situations to which game theory might apply, namely conflicts, seldom fulfill these requirements.

Still we have not touched on the greatest obstacle to casting conflicts in the framework of games of strategy. We have all along assumed that the game to be solved is a zero-sum game. If it is, one needs only to assign one's own utilities to the outcomes. In a nonzero-sum game, however, one must also assign the other's utilities to outcomes. How is this done? It hardly seems practical to ask the opponent how he values the various outcomes. One might well get false answers if the opponent thinks that it is to his advantage to deceive. And even if the opponent is willing to give the information, he is likely to be as arbitrary in his choices of utilities as we are, and so the uncertainties inherent in the problem multiply. If the opponent is reticent about giving information about his utilities, he will be all the more reticent about listing the strategies open to him. In military operations, it is precisely the *available strategies* which are the most jealously guarded secrets. Without knowledge of all available strategies both to self and to the opponent, the game cannot be cast into matrix form.

Suppose, however, that all of these difficulties have been overcome, and the situation has been cast into a game model, and the game turns out to be nonzero-sum of the type discussed in Chapters 8 and 9. The question of what strategy one is to choose still remains open if the full significance of nonzero-sum theory is to be grasped, for reasons which we hope have become clear in our analysis. It is at this point that game theory fails as a "know-how" theory.

We have spelled out the severe and, in our opinion, insuperable limitations of game theory as a prescriptive theory of rational decision in conflict situations. Wherein, then, does its usefulness lie?

Game theory, we think, is useful in the same sense that any sophisticated theory is useful, namely as a generator of ideas. The reason it is difficult to explain the usefulness of game theory in the limited conventional context of "usefulness" is because the really fruitful ideas of game theory are not those that lead to a more complete control of environment (as is the case with the natural sciences), but rather in quite a different direction. It is primarily because the usefulness of science is conventionally evaluated by its environment-controlling potential that the (projected) uses of game theory are still imagined in terms of "know-how." We have argued that this is largely an illusion. But this does not invalidate the usefulness of game theory, if the meaning of "usefulness" is extended beyond the enhancement of environment control.

We shall argue that the usefulness of game theory is somewhat akin to the usefulness of psychology and also, incidentally, that the usefulness of psychology is not that of a "know-how" science, whatever the imagined uses of psychology may be. In its role as a know-how science, psychology is easily degraded. Mass behavior manipulated to enhance the manipulators' power is an easily imagined by-product of psychological know-how; but it is questionable whether any benefit accrues to the human race from activities of this sort. Indeed, when we speak here of the usefulness of science, we shall speak of its usefulness to man, not to particular interests of power-wielding coalitions. If the manipulative applications of psychology are discounted, there still remains a tremendous potential of psychology as a science which can enhance man's insight into himself. This value tends to be ignored in a climate where manipulative techniques are held to be the most valuable product of knowledge.

Like psychology, game theory can be a source of ideas which lead to insights—insights into the nature of conflict based on the interplay of decisions.

In this role, the purely formal game theory has provided a rigorous logical basis of analysis of all forms of such conflict. If the insight-generating role of game theory is to be further developed, the next step is toward a descriptive theory, not a prescriptive theory. In my opinion, the prescriptive aspect of game theory ought to be written off for the following reasons: (1) game theory is inadequate as a theory of rational decision in the general context of nonnegotiable nonzero-sum games (which represent the overwhelming majority of real life conflicts among human beings); and (2) in the context of a zero-sum game, where game theory *is* an adequate theory of rational decision, its application is ultimately self-defeating. For if one side can find optimal strategies in a zero-sum situation, so can the other. No one gains from this "advance."

Consider for a moment the possibility of applying game theory to business competition. The competition for the share of the market (as distinguished from the size of the market) is by definition a zero-sum game.

There is no question that the use of effective game-theoretical techniques by some firms will instigate almost immediately the use of the same techniques by other firms. The total effect at first is bound to be simply the sharpening of competition. In the context of business competition, this effect can sometimes be rationalized as beneficial to the public at large (although I am not at all prepared to admit this in general). However, if we follow the logic of the zero-sum game to completion, we must conclude that the ultimate effect will be the *elimination* of competition. For the outcome of a zero-sum game is determined immediately, in the case of games with saddle points, and in the long run in the case of games without. A really complete game-theoretical analysis of a projected competitive game among firms ought to reveal how, for example, the market is to be divided between them, and to show to everyone's satisfaction that there is no point whatsoever in playing the game out. This may be a fantasy from the practical point of view, but this is where the "know-how" uses of game theory point.

While in the business sphere one can still make an argument about possible beneficent effects of competition, it is next to impossible to do so in the military context. Unless we espouse the point of view that the militarily strongest nations are the most worthy to survive, it is impossible to see what benefits derive from "improvements" of military technology. The same is true of "improvements" of military strategy, which are the chief reasons for game theory as seen by military professionals. No science can be kept a secret. Strategic improvements achieved by one military power can be immediately achieved by others. If this process led to actual definitive solutions of military conflicts viewed as zero-sum games (i.e., solutions on paper), this would indeed be an achievement of the first magnitude. No battle would need to be fought, no campaign waged, no war started. The outcomes would be clear in advance, and the power relations among states could be adjusted accordingly. It is, however, quixotic to suppose that these are the goals of the military profession. It is more realistic to suppose that the military professionals are as committed to their purely professional goals (i.e., to technological and strategic excellence) as the practitioners of any other profession. If, then, one believes that from the point of the human race (as distinct from the interests of power-wielding groups) the military profession contributes to misery rather than to well-being, one must necessarily hold the view that anything which enhances the effectiveness or the prestige of that profession is to be deplored rather than welcomed. The reason it is unnecessary to damn the theory of games as such on that account is that this theory is extremely ineffective as a prescriptive theory in actual application to real conflict situations, for reasons we have already pointed out.

Let us return to what we believe to be the creative, positive, and beneficent role of game theory. We have already mentioned the power of logical analysis provided by the purely formal theory of games, which gives us insights into the *logic* of strategic conflict, in particular the way it leads to the altogether novel idea that the very concept of rationality dissolves into ambiguities in certain conflicts not strictly competitive (nonzero-sum games). We also said that the formal theory could serve as the conceptual point of departure for a descriptive (empirical) theory of conflict. The aim of such a theory would be not to prescribe how people ought to conduct a conflict, but to describe how people do conduct themselves in conflicts.

Game theory, as it was formulated by mathematicians, is not equipped to deal with these matters, because there is no room in that theory for the psychological make-up of the participants. To the extent that psychological matters are allowed to enter a theory of conflict, the theory ceases to be a model of rational conflict. Its mathematical apparatus must then include parameters, so that conflict behavior would depend on these parameters. The theory would become a behavioral theory, and real behavior can never be explained on the basis of concepts of "rationality" alone. At least "rationality" must be modified to a relative concept to be put into specific psychological contexts.

For example, in our interactive model of game behavior (see Chapter 10), we assumed that the rate of change of the frequency with which a strategy will be chosen is proportional to the rate of change of the expected payoff with respect to that frequency. This was offered as a possible rationality principle. But the constant of proportionality appears in that model as a parameter. It may be different in different players, or in different classes of players, or it may vary with the payoffs of the game.

Thus the model we have proposed does not specify how a game will be played nor how it ought to be played. The model is an attempt to establish relations among the payoffs, the psychological parameters, and the behavior of players. The model may or may not be a good representation of what goes on, but in either case, *it has a theoretical leverage.* By comparing observations (usually obtained in controlled experiments) with the predictions of the model (after suitable values of the parameters have been put into the equations), we can form an idea about possible directions for further development of the theory.

The most important role of a model of this sort is not so much to describe the observations as to furnish interpretations of the parameters which fit the observations. To illustrate this role of the mathematical model, let us suppose for the moment that the variables x, y, z, and w (see pp. 152-53) do not change in the course of repeated plays of Game 39. It is, then, these quantities which constitute the parameters of the model. Suppose that by a proper assignment of values to these parameters we are able to get good agreement between some collection of experimental data and the predictions of the model. In view of the way the parameters have been defined, we can now proceed to interpret them in psychological terms.

Consider first the conditional propensity x. This propensity, we recall, is the probability that a player, following the outcome (CC), will again choose strategy C on the next play. Let us see whether this definition suggests a psychological interpretation of the parameter x. We may arrive at such an interpretation if we inquire into the psychological (instead of the strategic) structure of Game 39. We have seen that in this game, it is in the individual strategic interest of each player to choose strategy D, which dominates strategy C. Nevertheless the outcome (DD) is poor for both players, and this leads to the paradox that the choice of "rational" strategies leads to an undesirable result. We now ask, how is it possible to rationalize the choice of strategy C which, if chosen by both players, leads to a result desired by both players (CC).

To see how this can be done, let us examine the psychological rather than the strategic aspects of the situation depicted by Game 39. A player may choose strategy C in preference to D if (1) he trusts the other player to do the same, and (2) is himself trustworthy, i.e., does not succumb to the temptation to choose D (while the other chooses C) in order to get a bigger payoff. If the last outcome in a series of plays was (CC), he has some reason for trusting the other. Therefore the main motivation for sticking to C when (CC) has just occurred is something akin to trustworthiness. This may then be taken as an interpretation of the propensity x: trustworthiness.

Let us now consider the possible meaning of the propensity y. We recall that y is the probability that a player will choose C following a play on which he chose C and received a negative payoff as a result (because the other player chose D). The player, basing his estimate of what the other will do next time on what the other did last time, may well assume that the other will again choose D, especially since he was rewarded for that choice. If, therefore, in spite of this expectation, the "betrayed" player still chooses C, he is behaving either like a martyr or a fool, depending on the world view of whoever is evaluating this behavior. At any rate, we could associate the propensity y with a tendency to "forgive," or to try to elicit cooperation by example, or something of this sort. It is certainly a propensity to pursue a "soft" policy.

Next, consider the propensity z. This, we recall, is the tendency to play C after one has played D and been rewarded (because the other has played C). Here we have a tendency to exploit the good will of the other, or perhaps ordinary greed, since D coupled with the other's C brings in the biggest payoff.

Finally, consider the propensity w. This is the tendency to play C following a (DD) outcome. This tendency is akin to that represented by y, but perhaps not so much tinged with "martyrdom," because one might suppose that the other has played D "in self-defense" following one's own choice of D. Therefore, the shift to

C represents not so much a tendency to cooperate in the face of the other's clearly exploitative behavior as a readiness to initiate cooperation in a deadlock of mutual distrust.

We see that the interpretation of these parameters has considerable psychological richness. It is, of course, an open question to what extent people's behavior in a game of this sort (which in an experimental situation is usually played for trivial stakes) reflects actual psychological propensities. However, the translation from *in vitro* to *in vivo* is a problem in all experimental sciences. The fact remains that in the context of the experiment itself, these psychological interpretations can be suggestive. Next, note that although the psychological factors presumably operating touch on rather deep facets of character (trust, suspicion, trustworthiness, betrayal, forgiveness, repentance, etc.), nevertheless the measurements of these propensities is a perfectly straightforward matter. It requires no scaling, no arbitrarily selected indices. A frequency (which is what each of the propensities is) is a pure number—a fraction of the time a particular response is chosen. Mathematical models are most powerful when variables can be quantified in dimensionless units.[39]

Another source of new concepts which emerges from the analysis of nonzero-sum games is the multi-ordinal meaning of rationality. We have already seen how the seemingly clear notion of rationality (in the context of strategic decisions) must be separated into individual and collective rationality if the paradoxes immanent in some nonzero-sum games are ever to be resolved. The inductive (or dynamic) approach to game theory provides us with a further scale of refinement. We have seen that if two automata play Prisoner's Dilemma (see p. 151), independently adjusting the probability of the C response so as to maximize the long run expected payoff, they end up in the noncooperative trap. However, if they adjust not the direct probabilities of the C response but the *conditional* probabilities, the trend toward the noncooperative equilibrium is no longer inevitable. The dynamics of this process has an unstable equilibrium. Whether the automata will end up cooperating or not depends on the direction in which a deviation from this equilibrium takes place.

Adjusting conditional response probabilities can be taken as symptomatic of a higher order of "rationality" than adjusting direct probabilities. In common parlance, we often associate intelligence with a degree of conditionality of response (i.e., a degree of discernment of one's environment). We are thus led away from a "static" conception of rationality, which has dominated game theory, namely a rationality based on a complete knowledge of an *existing* state of affairs coupled with superhuman powers of deduction. We become aware of another "dynamic" conception of rationality, namely an ability to read the environment, to change one's hypotheses in accordance with acquired information and, by acting on the hypotheses, to *affect* the environment, in particular the perceptions of other actors like oneself. The so-called self-fulfilling assumption becomes in this context not merely a philosophically recognized possibility, but a genuine instrument of rational decision.

Finally, the theory of the negotiated game reveals the multi-ordinal meaning of equity. For example, Shapley, Nash, Raiffa, and Braithwaite have come up with different methods of arriving at a negotiated solution of a nonzero-sum game (see Chapter 8). One feels that there must be differences in the conceptions of the four men of what is "rational" or "equitable" or both. On the other hand, J. C. Harsanyi (1962) argues that one ought to distinguish solutions based on bargaining principles from those based on arbitration principles, in that only the latter involve the concept of equity. Without a precise definition of "equity," it is difficult to agree or disagree with this idea. In some cases it seems to be reasonable. Compare, for example, Shapley's and Nash's solutions of the two-person negotiated game. They differ only in their choice of the status quo point. Shapley chooses the point determined by the security levels of the players. Implicit in this choice is the idea that a person's share in the negotiated solution ought to reflect how much he effects improvement in his opponent's payoff over what his opponent would have gotten had he played the game as a zero-sum game, i.e., against a malevolent opponent. Hence a player gets the less the more he needs the cooperation of the other. This may be interpreted as an equity principle.

In Nash's solution, on the other hand, the status quo point is determined by the threat potentials of the players. The player gets the more, the more pressure he can bring to bear on the other. This is more difficult to interpret as an equity principle.

Although Raiffa's method of normalizing the utility scales appears as an equity principle, we see that it bears a strong resemblance to Nash's method in that both depend on the solution of a certain zero-sum game, in which the payoffs are differences of the payoffs in the original game, properly normalized.

Thus game-theoretical analysis directs a spotlight, as it were, on the assumptions which distinguish one principle of "equity" from another. (If one broadens the concept of equity to include "might is right," then the distinction between bargaining and arbitration disappears.) Game-theoretical analysis makes it possible to reduce disputes about equity to the really fundamental differences and so to avoid essentially unresolvable disputes about what, if anything, can be meant by "the greatest good for the greatest number," "fairness," "bargaining advantage," and similar concepts which, although certainly not entirely devoid of sense, can be different things to different people.

This leads us to the possible promise which the theory of the negotiated game holds out as an instrument of conflict resolution.

Ordinarily, one thinks only of nonzero-sum games as negotiable. It would seem that since in the zero-sum games the interests of the players are diametrically opposed, there is nothing to negotiate. However, in the larger context of negotiation, namely that of joint analysis of the situation, zero-sum games can also be "negotiated."

As an example, imagine a situation in a game of Chess in which one player clearly sees that he must win. If his opponent is a sufficiently competent Chess player, then he, too, will agree that inevitably victory must go to the first player. At this point it seems futile to continue the game.

In the days when war was the "sport of kings," negotiated settlements were the rule rather than the exception. Often the strategic potentialities open to the contending forces were taken into account in the negotiations. Even in our days of total and ideological wars, demands for surrender are sometimes coupled with appeals to humanitarian rationality, e.g., "to avoid useless bloodshed." These appeals are bids to recognize the nonzero-sum character of the conflict. If defeat of one side is inevitable, it often seems (especially to the prospective victor) that both sides ought to take advantage of a rational appraisal of the outcome and to avoid useless losses.

Torts offer, perhaps, even better examples of situations of this sort. Settlements out of court are results of recognition by the contending parties (or, rather by their attorneys, who are experienced in such matters) wherein lie the strengths and weaknesses of each side. These strengths and weaknesses may stem from the legal aspects of the case; and they may be economic, stemming from the resources at the disposal of each side for fighting the case through the courts.

Can the theory of the negotiated game ever be of help in settling torts, labor-management disputes, etc.? Again we must make a sharp distinction between the immediate application potential of the theory and its heuristic, concept-generating value. Our example of the Inspector-Evader game was designed to emphasize this difference. If our treatment of the situation appeared grotesque to the point of being a caricature, perhaps even a hoax, we must admit that this is exactly the way we meant it to appear. To take the various "solutions" seriously (including the negotiated settlements where Inspector deliberately refrains from inspection in order to allow Evader to evade, so that both can get larger payoffs) is to disregard completely the political, historical, psychological, and psychopathological underpinnings of the fundamental problems that have created the situation in the first place. Not that farces of the sort proposed are unknown. One is reminded of the strictly formalized duels of a century ago, in which gentlemen went through the motions of saving their "honor," at the same time taking elaborate precautions against serious consequences. A similar formalization of the Evader-Inspector "settlement" is certainly not inconceivable. However, it would be folly to identify such settlements with "rational conflict resolution." Genuinely rational conflict resolution demands an inquiry into the genesis of conflict. And this is precisely what game theory, with its marvelously sophisticated but utterly "blind" apparatus of analysis, completely ignores. The sine qua non of game theory is that it can get started only after the utilities are given. It never questions the rationality of the goals pursued by the contending parties.

There is, nevertheless, a lesson to be derived, if one *deliberately* forgets what the "interests" of the Evaders and the Inspectors reflect, and the probable consequences of pursuing them. If one views the end-result of the analysis not as a prescription to both the Inspectors and to the Evaders of what they ought to do, but rather a disclosure of the rich ramifications of the problem, then the very grotesqueness of the "solutions" is instructive. Especially the multiplicity of the solutions should give rise to some sober thought. If the analysis teaches us anything, it is how even the most drastic "stripping down" of a conflict to the simplest conceivable level (for there is no conflict simpler than a 2×2 game) reveals a maze of assumptions which must be made in order to get anywhere with the analysis. There are many choice points in this maze, and each leads to a different "solution." The lesson to be derived is that many of our cherished notions about every problem having an "answer," about the existence of a "best" choice among a set of courses of action, about the power of rational analysis itself, must be relegated to the growing collection of shattered illusions. Rational analysis, for all its inadequacy, is indeed the best instrument of cognition we have. But it often is at its best when it reveals to us the nature of the situation we find ourselves in, even though it may have nothing to tell us how we ought to behave in this situation. Too much depends on our choice of values, criteria, notions of what is "rational," and, last but by no means least, the sort of relationship and communication we establish with the other parties of the "game." These choices have nothing to do with the particular game we are playing. They are not *strategic* choices, i.e., choices rationalized in terms of advantages they bestow on us in a particular conflict. Rather they are choices which we make because of the way we view ourselves, and the world, including the other players. The great philosophical value of game theory is in its power to reveal its own incompleteness. Game-theoretical analysis, if pursued to its completion, *perforce* leads us to consider other than strategic modes of thought.

Notes

1. To the layman the presence or absence of chance events seems important in classifying games, as reflected in the distinction between "games of chance" and "games of skill." From the point of game theory, however, the important distinction is not the interventions of chance but the information about the *results* of the intervention. If the results are known (as, for example, in dice games) one theory applies; but if they are not (as in card games), additional concepts must be introduced. In particular, in the latter case, it is often to a player's advantage to allow chance to decide his own choices (see Chapter 6).

2. Games with only one bona fide player are sometimes called "games against nature" or "one person games." Gambling theory, decisions determined by actuarial principles, and many problems treated in what is known as "operations research" ("operational research" in Britain) can be viewed as falling within the scope of one person game theory.

3. In the early writings on probability theory the expected utility was sometimes called "moral expectation." This was essentially the mean of all possible utilities with respect to their probabilities of occurrence.

4. I feel that this point must be strongly stressed because of the constant temptation to view game theory as a source of techniques for a rational conduct of conflicts.

5. Consistency of choice implies, among other things, a transitive preference relation. Thus if a player prefers Ticket #1 to

Ticket #2 and the latter to some other Ticket #3, then to be consistent, the player must prefer Ticket #1 to Ticket #3. When the prizes are numerous and diverse such consistency can by no means be taken for granted.

6. For example, if an outcome is "worth" 1 utile on one party's utility scale, and 2 utiles on the other's, we can by no means say that the outcome is worth more to the second party than to the first; for by choosing different units or different zero points we can easily reverse the inequality.

7. The square of the length of a diagonal of a rectangular parallelopiped (a box) equals the sum of the squares of its three dimensions.

8. If a quantity is at least as great as —v, its negative does not exceed v. For convenience we refer all quantities to v rather than to —v.

9. In mathematics "degeneracy" refers to a simplification resulting from a limiting case. For example, consider a rectangular parallelopiped one of whose dimensions keeps shrinking. When this dimension becomes zero, the parallelopiped changes into a rectangle. Thus a rectangle can be viewed as a "degenerate parallelopiped." Similarly, by varying a payoff of a game without a saddle point we can change it to one with a saddle point. Since from the point of view of game theory, games with saddle points are simpler than those without, the former can be viewed as "degenerate cases" of the latter.

10. A sub-game of a game is a game whose payoff matrix is obtained by striking some rows and/or columns from the matrix of the original game.

11. Note that the phrase "in the same degree" implies interpersonal comparison of utilities, hitherto excluded from our considerations. In a zero-sum game, even if the players can pool their payoffs (which may be, for example, money) to be apportioned between them by a negotiated settlement, there can be no outcome which is preferred by *both* to another outcome.

12. Again an interpersonal comparison of utilities is implied. It is, perhaps, inevitable in negotiations.

13. Some authors, notably J. Harsanyi (1962), make a distinction between bargaining and arbitration. The latter presumably makes use of some equity principles based on some social norms while the former does not. However, both bargaining and arbitration presuppose some principles of rationality. Since I believe that notions of what is "rational" (in situations depicted as nonzero-sum games) are also socially determined, I do not see as

sharp a distinction between bargaining and arbitration as does Harsanyi.

14. A convex polygon is one all of whose angles have less than 180 degrees. More generally, if two points inside or on the boundary of a "convex body" are joined by a straight line, every point in that line lies inside or on the boundary of the body.

15. This expectation is not justified on psychological grounds. It is simply a consequence of assuming utilities to be measured on an interval scale. This assumption is retained in game theory whenever possible.

16. The derivative of a quantity is essentially its rate of change with respect to another quantity. If this rate of change alters continuously, then it must vanish whenever a maximal (or minimal) value is reached.

17. It is an open question whether this reference of all utility scales to a "standard" scale does not in itself imply a subtle introduction of interpersonal comparison of utilities.

18. Note that in a zero-sum game the maximin and the minimax strategies are identical.

19. The two normalizations reflect two different "equity principles."

20. Note that the payoff of -2 to Pollux at (C_2, P_1) is actually not a threat at Castor's disposal; for if Castor "threatens" C_2, Pollux has an excellent counter-threat, P_2, which gives him the largest payoff of $+2$, at the same time punishing Castor with -2.

21. If the game is played several times, psychological complications arise. Pollux can decide to take the punishment in order to "teach Castor a lesson." Since players' learning capacities do not fall within the scope of game theory proper, this possibility cannot be discussed in the present context.

22. This argument, like others based on the "prominence" of a particular outcome, presupposes psychological characteristics of the players, i.e., carries the discussion beyond game theory proper. In particular, the prominence of (C_2, P_2) depends on an *absolute* utility scale. It is lost if the utilities of one player undergo a linear transformation.

23. The game has been studied in experimental contexts by several investigators. As an example, see A. Rapoport and A. Chammah (1965). A review of the literature on this game is by P. S. Gallo and C. McClintock (1965).

24. In Schelling's view (conveyed by personal communication), a "prominent outcome" must be an equilibrium. Thus the outcome (C_1, C_2) does not qualify as a "prominent" outcome. If,

however, the equilibrium requirement is dropped, (C_1, C_2) certainly appears as a prominent outcome, since it is the only outcome in which both players "win" (assuming an absolute utility scale).

25. Note that if the game is represented by a strategy matrix, simultaneous or independent choice of strategies is assumed; thus it is impossible from the point of view of game theory to "communicate intentions" in this context.

26. Clearly this is my own judgment. However, it has been repeatedly reinforced in conversations with other game theorists.

27. M. M. Flood introduced the term game-learning theory to encompass these ideas. For an example of this approach, see P. Suppes and R. C. Atkinson (1960).

28. A differential equation relates variables and their rates of change with respect to each other. A solution of a differential equation is a relation among the variables only, from which the original relations can be deduced. For example, the statement that the rate of growth of a population is proportional to the population's size is expressed as a differential equation. A solution of this differential equation expresses the size of the population as an exponential function of time.

29. Consider two differential equations in which variables x and y and their rates of change with respect to time t are related. A solution is a pair of expressions giving x and y as functions of time. In the phase space, x and y are plotted against each other. Thus the phase space exhibits the relations which x and y have to each other but not to time. Viewing it another way, a solution may be thought of as the motion of a point (with variable coordinates x and y) in the phase space. The track or orbit of the point is the plot of y against x in phase space.

30. This result is obtained by integrating (summing) the time functions of x and y over long time intervals and dividing by the length of these intervals.

31. Game 38 will be recognized as a general form of Prisoner's Dilemma.

32. The expected payoffs are obtained by averaging the possible payoffs weighted by the probabilities of their occurrence. The latter one occurred as the asymptotic solution of the stochastic process which constitutes our present model.

33. This formulation was proposed by H. Raiffa (personal communication).

34. For example, one can assume a cost associated with inspections. Hence in the case of no evasion, no inspection is preferred to inspection.

35. I.e., Evader can guarantee himself this amount.

36. For the rationale of this procedure, see for example, R. D. Luce and H. Raiffa, *Games and Decisions,* chapter 6.

37. This is the most outspoken form of collusion between Evader and Inspector. We have already raised the question of "realism" of the negotiated game model. If the frank collusion seems unrealistic, we suspect it is because of the semantic connotations of the terms "Evader" and "Inspector." In reality, the feasibility of the collusion reflects only the fact that to a certain extent the interests of the two players coincide.

38. See discussion of Nash's generalized bargaining problem in chapter 8.

39. Generally speaking, the stronger the scale, the less restricted is the class of mathematical models in which the corresponding quantities can enter as variables. For a discussion of this principle see R. D. Luce (1959).

References

Bibliographies on game theory and related topics are included in most books on the subject. The following references are suggested to the reader who wishes to familiarize himself more thoroughly with the topics selected in this book.

Books on Game Theory

Williams, J. D. *The Compleat Strategyst.* New York: McGraw-Hill, 1954. (An elementary discussion of the two-person zero-sum game.)

Rapoport, A. *Fights, Games, and Debates.* Ann Arbor: The University of Michigan Press, 1960. (Part II is an elementary exposition of game theory.)

Luce, R. D. and Raiffa, H. *Games and Decisions.* New York: John Wiley & Sons, 1957. (A rather complete exposition with emphasis on social science applications. Contains a full bibliography of the literature through 1955. Presupposes some mathematical background.)

Von Neumann, J. and Morgenstern, O. *Theory of Games and Economic Behavior.* New York: John Wiley & Sons, 3rd ed., paper, 1964, Science Editions. (The original treatise. Presupposes an extensive mathematical background.)

Discussions of Nonzero-game Theory

Braithwaite, R. B. *Theory of Games as a Tool for the Moral Philosopher.* Cambridge: Cambridge University Press, 1955.

Harsanyi, J. C. "On the rationality postulates underlying the theory of cooperative games." *Journal of Conflict Resolution,* 5 (1961), 179-96.

———. "Rationality postulates for bargaining solutions in cooperative and noncooperative games." *Management Science,* 9 (1962), 141-53.

Nash, J. F. "The bargaining problem." *Econometrica,* 18 (1950), 155-62.

———. "Noncooperative games." *Annals of Mathematics,* 54 (1951), 286-95.

———. "Two-person cooperative games." *Econometrica,* 21 (1953), 128-40.

Raiffa, H. "Arbitration schemes for generalized two-person games," in Kuhn, H. W. and Tucker, A. W. (eds), *Contributions to the Theory of Games,* II (Annals of Mathematics Studies, 28), Princeton, N.J.: Princeton University Press, 1953.

Schelling, T. C. *The Strategy of Conflict.* Cambridge, Mass.: Harvard University Press, 1960.

Shapley, L. S. "A value for n-person games," in Kuhn, H. W. and Tucker, A. W. (eds), *Contributions to the Theory of Games,* II (Annals of Mathematics Studies, 28). Princeton, N.J.: Princeton University Press, 1953.

Monographs on Gaming Experiments

Rapoport, A. and Chammah, A. M. *Prisoner's Dilemma: A Study of Conflict and Cooperation.* Ann Arbor: The University of Michigan Press, 1965.

Suppes, P. and Atkinson, R. C. *Markov Learning Models for Multiperson Interactions.* Stanford: Stanford University Press, 1960.

Discussion of the Relation between Scales of Measurement and Mathematical Models

Luce, R. D. "On the possible psychophysical laws." *Psychological Review,* 66 (1959), 81-95.

Game Theory, Gaming, and Applications

Shubik, M., (ed.). *Game Theory and Related Approaches to Social Behavior.* New York: John Wiley & Sons, paper, 1964. (Introductory exposition of game theory, essays on application, gaming experiments.)

Index

(Numbers marked n refer to the Notes; those marked b refer to the References)

N-PERSON GAME THEORY: CONCEPTS AND APPLICATIONS *by Anatol Rapoport*

SOIL ANIMALS *by Friedrich Schaller*

THE EVOLUTION OF MAN *by G.H.R. von Koenigswald*

SPACE CHEMISTRY *by Paul W. Merrill*

THE LANGUAGE OF MATHEMATICS *by M. Evans Munroe*

CRUSTACEANS *by Waldo L. Schmitt*

TWO-PERSON GAME THEORY: THE ESSENTIAL IDEAS *by Anatol Rapoport*

THE ANTS *by Wilhelm Goetsch*

THE STARS *by W. Kruse and W. Dieckvoss*

LIGHT: VISIBLE AND INVISIBLE *by Eduard Ruechardt*

THE SENSES *by Wolfgang von Buddenbrock*

EBB AND FLOW: THE TIDES OF EARTH, AIR, AND WATER *by Albert Defant*

THE BIRDS *by Oskar and Katharina Heinroth*

PLANET EARTH *by Karl Stumpff*

ANIMAL CAMOUFLAGE *by Adolf Portmann*

THE SUN *by Karl Kiepenheuer*

VIRUS *by Wolfhard Weidel*

THE WASPS *by Howard Evans and Mary Jane Eberhard*